Blood - Updates on Hemodynamics and Thalassemia

Edited by Aise Seda Artis

Published in London, United Kingdom

IntechOpen

Supporting open minds since 2005

Blood - Updates on Hemodynamics and Thalassemia
http://dx.doi.org/10.5772/intechopen.95185
Edited by Aise Seda Artis

Contributors
Nitu Nigam, Prithvi Kumar Singh, Suhasini Bhatnagar, Sanjay Kumar Nigam, Anil Kumar Tripathi, Ahmed Mutlaq Shemran Alwataify, Sabih Salih Alfatlawy, Yahia Abid Alshahid Altufaily, Iman El-Baraky, Julio Garcia, Patrick Geeraert, Hansuk Kim, Safia Ihsan Ali, Shirin Aliabadi, Ashifa Hudani, Hourieh Jamalidinan, Monisha Ghosh Srabanti, Ildar I. Lutfarakhmanov, Petr I. Mironov, Ildar R. Galeev, Valentine N. Pavlov, Nakul Ravikumar, William T. McGee, Geoffrey R. Sheinfeld, Kenneth J. Dormer, John M. Owen, Aise Seda Artis

First published in London, United Kingdom, 2022 by IntechOpen
IntechOpen is the global imprint of INTECHOPEN LIMITED, registered in England and Wales, registration number: 11086078, 5 Princes Gate Court, London, SW7 2QJ, United Kingdom
Printed in Croatia

British Library Cataloguing-in-Publication Data
A catalogue record for this book is available from the British Library

Additional hard and PDF copies can be obtained from orders@intechopen.com

Blood - Updates on Hemodynamics and Thalassemia
Edited by Aise Seda Artis
p. cm.
Print ISBN 978-1-83969-716-6
Online ISBN 978-1-83969-717-3
eBook (PDF) ISBN 978-1-83969-718-0

For EU product safety concerns:
IN TECH d.o.o., Prolaz Marije Krucifikse Kozulić 3, 51000 Rijeka, Croatia, info@intechopen.com or visit our website at intechopen.com.

Meet the editor

Aise Seda Artis, MD, is an Associate Professor of Physiology at Western Balkans University, Tirana, Albania. She graduated from Istanbul University Cerrahpasa School of Medicine, Istanbul, Turkey, in 1998. She has experience working as a general practitioner and researcher at clinics in Turkey and the United States. During her training in physiology at Erciyes University School of Medicine, Kayseri, Turkey, Dr. Artis was mainly involved in hemorheological and neuroscientific research. After training, she worked in the same department as an academic staff member. Prior to her current position, she worked at Istanbul Medeniyet University School of Medicine, Istanbul, Turkey.

Contents

Preface

Blood is a type of connective tissue in fluid form. Formed elements suspended in the plasma fluid circulate throughout the body within the cardiovascular system. This book examines both the fluid and cellular components of this fascinating liquid.

Following an introductory chapter, the book begins with a section dedicated to novel developments and updates on various topics in hemodynamics.

A classical area of physiology, namely hemodynamics, explains the physical laws that govern the flow of blood in the vessels. The hemodynamic response continuously monitors and adjusts to conditions in the body and its environment. Understanding the interaction between different blood flow variables helps to better interpret their implications in various physiopathological conditions. Future research is expected to focus mainly on the relations between hemodynamics and biological processes with active cellular responses.

Chapter 2, "Hemodynamic Perspectives in Anemia", discusses hemodynamic perspectives in anemia patients in critical condition. Emphasizing the transfusion threshold and significance of oxygen delivery and consumption, it shares new modalities to assess volume status and presents the algorithm guiding resuscitation.

Chapter 3, "Four-Dimensional Flow Magnetic Resonance Imaging and Applications in Cardiology", discusses data acquisition and pre-processing in four-dimensional flow magnetic resonance imaging (4D flow MRI) and its cardiologic applications. It gives insight into the use of 4D flow MRI in congenital heart disease, mitral regurgitation, atrial fibrillation, and different aortic pathologies like stenosis, coarctation, and bicuspid valve disease.

Chapter 4, "Cardiovascular Changes during Robot-Assisted Pelvic Surgery", elaborates cardiovascular complications of robot-assisted laparoscopic pelvic surgery. It highlights the underlying pathophysiology and the importance of perioperative monitoring.

Chapter 5, "Hemodynamic Alterations in Multiple Sclerosis", presents the cervical and cerebral changes and possible mechanisms of altered perfusion in multiple sclerosis. It also shares methods for the evaluation of perfusion and the clinical implications of these alterations.

Finally, Chapter 6, "The Shear Stress/KLF2/Nrf2/ARE Pathway: A Hemodynamic Defense against Oxidative Stress", discusses the importance of hemodynamic laminar shear stress in oxidative homeostasis and how to overcome oxidative stress through hemodynamics by the induced shear stress/Krupple-like factor2/Nrf2/ARE pathway.

The second section of the book shares novel developments in thalassemia.

The thalassemias are related to red blood cells. Among other hemoglobinopathies, they represent a major group. Abnormal hemoglobin formation is the result of inherited disorders of hemoglobin synthesis. The two main categories of the disease are alpha and beta-thalassemia, which are then divided into further subcategories. Some patients with mild forms of the disease might even go unnoticed and present with only mild anemia and iron deficiency problems. However, other more severe forms can even result in death. Complications related to compromised immunity and cardiovascular issues require a cautious clinical approach in thalassemia.

Chapter 7, "An Early Diagnosis of Thalassemia: A Boon to a Healthy Society", shares diagnostic strategies and screening methods for thalassemia with emphasis on cost comparison and point-of-care testing.

Chapter 8, "Pulmonary Hypertension in Thalassemia Patients", reviews in detail pulmonary hypertension, which is an important cause of morbidity and mortality. This progressive condition is one of the most common cardiac complications. The chapter discusses every aspect from etiology to treatment and recommendations.

Finally, Chapter 9, "Challenges of Hepatitis C Virus Treatment in Thalassemia", discusses the approach to thalassemia patients with hepatitis C infection. Thalassemia patients are especially predisposed to multi-transfusion-acquired hepatitis C virus. The chapter presents novel treatments and pharmacokinetic considerations.

Aise Seda Artis
Western Balkans University,
Tirana, Albania

Section 1

Introduction

Chapter 1

Introductory Chapter: Fascinating Blood

Aise Seda Artis

1. Introduction

"The blood jet is poetry and there is no stopping it."

Sylvia Plath, 'Ariel'.

Ancient people were aware of blood's importance and fascinated by its mystery. According to the belief, Medusa had two kinds of blood circulating through her vessels: lethal blood on her left side, and life-giving blood on her right side. Later the humoral pathology theory of Hippocrates proposed the human body as a vessel for four liquids: yellow bile, black bile, white phlegm, and red blood. Each corresponds to one of the four classical elements – fire, earth, water, and air and one of the traditional four temperaments – sanguine, choleric, phlegmatic, and melancholic. Later, Galen proposed that blood was made in the liver from food and drink carried from the digestive tract. An ideal balance among the four humors was the key to good health [1]. Inspecting blood samples from patients and determining the relative amounts of each humor was the method for the diagnosis of an imbalance. Bloodletting treatments were common practice, and described in detail by Avicenna in his "Canon of Medicine". This theory shaped the practice in Greek, Roman, and Islamic philosophy and medicine for many centuries until the nineteenth century [2].

In today's view, blood is a type of connective tissue in fluid form. Formed elements suspended in the plasma fluid circulate throughout the body within the cardiovascular system. While the primary function is to transport oxygen, nutrients, and other substances; its specific functions also include defense, distribution of heat, and maintenance of homeostasis throughout the body.

2. Hemodynamics

Hemodynamics is the study of the mechanical and physiologic properties controlling blood pressure and flow through the body, "the physical study of flowing blood and of all the solid structures (such as arteries) through which it flows" [3]. This classical area of physiology emphasizes the fluid and solid mechanics of the cardiovascular system, concerning the distribution of pressures and flows in the circulatory system. Control of the circulatory system is through the homeostatic mechanisms of autoregulation. Many biological processes, physical factors influence blood flow (and vice versa). In addition to systemic hemodynamic alterations, microvascular alterations are frequently observed in critically ill patients. The hemodynamic response continuously monitors and adjusts to conditions in the body and its environment. Some parameters have been defined to quantify blood flow and its relationship with systemic circulatory changes. There is a constant development of the instrumentation and the techniques together with increasing capabilities for numerical computation. Complex and extensive factors influence hemodynamics. Understanding the interactions between different blood flow variables help to better interpret their implications in variable physiopathological conditions. Also, the help of improved imaging techniques is undeniable. Future research is expected to focus mainly on the interactions between hemodynamics and biological processes involving active cellular responses [4].

3. Hemorheology

The study of blood flow is called hemodynamics, and the study of the properties of the blood flow is called hemorheology. Blood is a non-Newtonian shear-thinning fluid and therefore is most efficiently studied using rheology rather than hydrodynamics [5]. The blood flow properties include blood viscosity which depends on plasma viscosity, hematocrit, red blood cell (RBC) aggregation, and RBC deformability [6]. The RBCs have a unique ability to deform and pass through small capillaries before rapidly recovering their initial shape. Under most flow conditions RBCs behave like fluid drops [7]. Due to the liquid-like behavior of RBCs under shear stress, blood can also be considered as a liquid-liquid emulsion. White blood cells and platelets can also affect blood rheology but, under normal conditions, RBCs represent most of the cellular components and make the biggest contribution to blood viscosity [8].

Hemorheological alterations occur in a wide range of physiological and pathophysiological conditions. In many diseases, the deformability of RBCs is impaired as a result of defects in cell membrane skeletal architecture, RBC aging, and mechanical damage [9–11]. Adverse hemorheological alterations may decrease tissue perfusion. Since control of blood flow is directly related to the metabolic conditions of the tissue, the perfusion change can be compensated by controlling the vascular geometry (i.e., diameter) component of flow resistance.

4. Thalassemia

RBCs show rheological abnormalities, also when thalassemia is present. Currently seen most frequently in the tropical belt, thalassemias remain a serious global health problem. Thalassemias are a group of inherited microcytic, hemolytic anemias characterized by defective hemoglobin synthesis. Alteration of the cell membrane may result from the interaction between the defective hemoglobin chain and the membrane cytoskeleton [12]. However, other changes are caused by hemoglobin denaturation [13]. Previous works suggest that the decreased RBC deformability is probably due to microcytosis that is present even in thalassemia carriers [14]. Recent work also suggests the presence of additional but yet unknown causes [15]. Thalassemia results from unbalanced hemoglobin synthesis caused by decreased production of at least one globin polypeptide chain. Symptoms and signs result from anemia, hemolysis, splenomegaly, bone marrow hyperplasia, and if there have been multiple transfusions, iron overload. Endothelial injury, splenomegaly, and transfusion-related hemodynamic alterations play an important role in the altered hemodynamics of thalassemia [16–18].

5. Conclusion

For future research, the concepts of hemodynamics and hemorheology will retain their importance with their basis in the understanding of fluid and solid mechanics. A better understanding of the biological processes involving active cellular responses provided by advanced techniques is the key.

To serve this purpose, the first section of the present book is dedicated to present novel developments and updates on various topics in the frame of hemodynamics:

- Four-dimensional flow magnetic resonance imaging (4D flow MRI) is a relatively new imaging technique. It offers the ability to measure and visualize the temporal evolution of complex blood flow patterns within an acquired 3D volume, allowing for the computation of multiple hemodynamic metrics.

- Robot-assisted laparoscopic pelvic surgery seems safe and efficacious. A thorough understanding of the underlying pathophysiology together with meticulous monitoring would help to better cope with the possible cardiovascular complications.

- Autoimmune diseases is another factor affecting hemodynamic homeostasis. Multiple sclerosis may present with altered cervical and cerebral perfusion that is important for the pathophysiology and the clinical implications of the disease.

- Recent work designates oxidative stress as a critical determinant of blood flow dynamics. Hemodynamic laminar shear stress in oxidative homeostasis regulation by the induced shear stress/Krupple-like factor2/Nrf2/ARE pathway appears significant.

- Anemia patients under critical conditions require hemodynamic guidance for accurate assessment of the volume and the transfusion threshold and deciding the best algorithm for the resuscitation of these patients.

The second section of this book is a thalassemia update. Concisely, this section gives an update and novel developments in thalassemia. Four chapters present an outline of B-thalassemia, the early diagnostic tools, and strategies in thalassemia, together with the approaches for the obstacles of pulmonary hypertension and Hepatitis C virus infection.

The chapters of this book may not seem to be in complete harmony. However, in the present era of fast communications, we aimed to provide some novelties in the field. This work serves an audience from different backgrounds providing a review on a variety of selected topics with the purpose of an update.

Acknowledgements

I would like to thank Prof Talip Asil for his kind effort in reviewing one of the chapters. I would also like to appreciate Author Service Manager Ms. Marica Novakovic and Commissioning Editor Ms. Ana Simcic for their constant help.

Author details

Aise Seda Artis[1,2]

1 Istanbul Medeniyet University, Istanbul, Turkey

2 Vistula University, Warsaw, Poland

*Address all correspondence to: aseda@yahoo.com

References

[1] Dammeyer J, Zettler I. A brief historical overview on links between personality and health. In: Personality and Disease. Amsterdam, Netherlands: Elsevier; 2018. pp. 1-16

[2] Imankulova T, Ysmailova R, Salieva D, Mamatova A, Mukhtarova N, Darbanov B, et al. Avicenna's contribution to medical terminology. OALib. 2020;**7**(9):1-9

[3] McDonald D. In: Arnold E, editor. Blood Flow in Arteries. London: Williams & Wilkins; 1974. pp. 459-488

[4] Secomb TW. Hemodynamics. In: Comprehensive Physiology. Hoboken, NJ: Wiley; 2016. pp. 975-1003

[5] Liu H, Lan L, Abrigo J, Ip HL, Soo Y, Zheng D, et al. Comparison of newtonian and non-newtonian fluid models in blood flow simulation in patients with intracranial arterial stenosis. Frontiers Physiology. 2021;**12**:718540

[6] Nader E, Skinner S, Romana M, Fort R, Lemonne N, Guillot N, et al. Blood rheology: Key parameters, impact on blood flow, role in sickle cell disease and effects of exercise. Frontiers Physiology. 2019;**10**:1329

[7] Schmid-Schönbein H, Wells RE, Goldstone J. Fluid drop-like behaviour of erythrocytes – Disturbance in pathology and its quantification. Biorheology. 1971;**7**(4):227-234

[8] Pop GAM, Duncker DJ, Gardien M, Vranckx P, Versluis S, Hasan D, et al. The clinical significance of whole blood viscosity in (cardio)vascular medicine. Netherlands Heart Journal. 2002; **10**(12):512-516

[9] Faustino V, Rodrigues RO, Pinho D, Costa E, Santos-Silva A, Miranda V, et al. A microfluidic deformability assessment of pathological red blood cells flowing in a hyperbolic converging microchannel. Micromachines. 2019; **10**(10):645

[10] Simmonds MJ, Meiselman HJ, Baskurt OK. Blood rheology and aging. Journal of Geriatric Cardiology. 2013;**10**(3):291-301

[11] Rico LG, Juncà J, Ward MD, Bradford JA, Bardina J, Petriz J. Acoustophoretic orientation of red blood cells for diagnosis of red cell health and pathology. Scientific Reports. 2018;**8**(1):15705

[12] Yuan J, Kannan R, Shinar E, Rachmilewitz EA, Low PS. Isolation, characterization, and immunoprecipitation studies of immune complexes from membranes of beta-thalassemic erythrocytes. Blood. 1992;**79**(11):3007-3013

[13] Shinar E, Rachmilewitz EA. Oxidative denaturation of red blood cells in thalassemia. Seminars in Hematology. 1990;**27**(1):70-82

[14] Vayá A, Iborra J, Falcó C, Moreno I, Bolufer P, Ferrando F, et al. Rheological behaviour of red blood cells in beta and deltabeta thalassemia trait. Clinical Hemorheology and Microcirculation. 2003;**28**(2):71-78

[15] Krishnevskaya E, Payán-Pernía S, Hernández-Rodríguez I, Remacha Sevilla ÁF, Ancochea Serra Á, Morales-Indiano C, et al. Distinguishing iron deficiency from beta-thalassemia trait by new generation ektacytometry. International Journal of Laboratory Hematology. 2021;**43**:e58-e60

[16] Butthep P, Nuchprayoon I, Futrakul N. Endothelial injury and altered hemodynamics in thalassemia. Clinical Hemorheology and Microcirculation. 2004;**31**(4):287-293

[17] Aessopos A, Farmakis D, Tsironi M, Deftereos S, Tassiopoulos S, Konstantopoulos K, et al. Hemodynamic assessment of splenomegaly in β-thalassemia patients undergoing splenectomy. Annals of Hematology. 2004;**83**(12):775-778

[18] Mut MA, Türkkan E, Dağ H, Dursun H. Evaluation of transfusion-related hemodynamic parameters in patients with beta-thalassemia major by ambulatory blood pressure monitoring method. Iberoamerican Journal of Medicine. 2021;**3**(3):187-195

Section 2

Hemodynamics

Chapter 2

Hemodynamic Perspectives in Anemia

Nakul Ravikumar, Geoffrey R. Sheinfeld and William T. McGee

Abstract

Oxygen delivery in normal physiologic states is determined by cardiac output, hemoglobin, oxygen saturation, and to a lesser extent, dissolved oxygen in the blood. Compensatory mechanisms such as an increase in stroke volume, heart rate, and re-distribution of blood flow helps in scenarios with increased oxygen demand. In cases of acute hemodynamic decompensation, this pre-existing physiologic relation between oxygen delivery and oxygen consumption is altered, resulting in tissue hypoxia and resultant anaerobic metabolism. A persistent state of sub-critical O_2 delivery correlates with increased mortality. Oxygen consumption itself is usually independent of delivery unless a critical threshold is unmet. We can use various parameters such as serum lactate, oxygen extraction, and central venous oxygen saturation to determine this pathology. A basic understanding of this physiology will help better tailor therapy to improve outcomes in critically ill patients.

Keywords: oxygen delivery (DO_2), oxygen consumption (VO_2), central venous oxygen saturation ($ScVO_2$), anemia, stroke volume variation (SVV), pulse pressure variation (PPV), passive leg raising (PLR), oxygen extraction (O_2ER)

1. Introduction

Anemia in critically ill patients is almost inevitable. The etiology of this can be varied, including overt blood loss, bone marrow suppression, functional iron deficiency, and decreased erythropoiesis [1]. Reports indicate that more than 95% of patients in an ICU are anemic by Day 8. Interventions such as scheduled daily phlebotomies lead to about 70-100 cc/day blood loss, while body turnover is about 15-20 cc/day. Little do we realize that this "anemia of chronic investigation" can lead up to 1/3 of transfusions when considering the current transfusion threshold practice of hemoglobin (Hb) less than 7 g/dL [2]. However, with numerous methods currently available to assess transfusions' physiological response, treating an arbitrary number is not the most refined approach. Parameters such as oxygen delivery (DO_2), oxygen consumption (VO_2), oxygen extraction ratio (O_2ER), and cardiac output are of paramount importance in the decision to transfuse RBCs.

This chapter aims to discuss the physiology of oxygen delivery, the pathophysiology of lung injury related to volume overload, and transfusion-related injury. Ultimately, we would like to demonstrate how employing a simple physiologic algorithm using parameters such as stroke volume, and stroke volume variability can lead to better care in this population.

IntechOpen

2. Transfusion threshold in critically ill patients

The trigger to order blood products in the medical ICU is often a reflex action. In critically ill patients, the goal of transfusion in a volume replenished patient should aim to improve oxygen delivery and thereby accomplish the end goal of improving oxygen uptake in tissues. There is much evidence regarding liberal vs. restrictive transfusion limits in various subsets of patients. Vastly discussed strategy being restrictive transfusion (less than 7 g/dL) vs. liberal transfusion goals (9-10 g/dL). We can all agree that 7 g/dL is presently the customary number across many diseases. The TRICC trial randomized 838 critically ill patients to liberal (goal hemoglobin 10–12 g/dL) versus restrictive (goal hemoglobin 7–9 g/dL) transfusion strategies observed no benefit to liberal transfusion. However, the study almost achieved ($p=0.11$) statistical significance for finding increased mortality in patients randomized to liberal transfusion. This trial showed that there was no 30-day mortality difference between restrictive strategy vs. liberal strategy. Still, it showed harm in the subset of younger patients with lower APACHE scores and clinically significant cardiac disease [3]. Current data suggests that restrictive strategy in critically ill patients decreases the number of transfusion reactions, length of stay, and mortality [4]. Various studies focusing on patients with ARDS, trauma, and sepsis have shown that tissue oxygenation and extraction does not improve with transfusion alone. This disparity is attributable to poor flow characteristics, high O_2 affinity secondary to reduced 2,3 DPG concentration in transfused blood and increased viscosity affecting functional capillary diameter. While large multicenter trials recently have shown a higher threshold to transfuse is beneficial or at least non-injurious, it is prudent to focus on the physiological response a patient may have to blood transfusion.

3. Importance of measuring DO_2 and VO_2

Does it make a difference? Each gram of Hemoglobin can carry 1.34 cc of O_2, with each molecule 100% saturated as it enters systemic circulation.

DO_2 is derived from the following (Eq. (1)):

$$DO_2 = \text{Cardiac output} \times \text{Arterial oxygen content}\left(CaO_2\right) ml / \text{min} / kg \quad (1)$$

$$CaO_2 = \left(1.34 \times Hb\right) \times SaO_2 + \left(0.003 \times PaO_2\right) ml / dl \quad (2)$$

$$\left(0.003 \times PaO_2 = \text{Dissolved oxygen in the blood}\right)$$

VO_2 is calculated by the difference between oxygen content in arterial blood and venous blood (Eq. (4)).

$$\text{Venous oxygen content}\left(CvO_2\right) = \left(1.34 \times Hb\right) \times SvO_2 + \left(0.003 \times PvO_2\right) ml / dl \quad (3)$$

Hence,

$$VO_2 = \text{Cardiac output} \times \left(1.34 \times Hb\right) \times \left(SaO_2 - SvO_2\right) \quad (4)$$

The fundamental goal of RBC transfusion is to improve oxygen content and to improve tissue oxygenation and delivery. While decreasing hemoglobin concentration can lead to critical oxygen content and delivery, this can be compensated by improving cardiac output and maintaining a euvolemic status.

VO_2 calculated with this equation is called the reverse Fick method. With an increase in oxygen demand, VO_2 increases from a baseline of 25%. Although there is no clear-cut point, oxygen extraction number between 33 and 40% is considered acceptable and not critical. If the oxygen extraction is high, emphasis should be on delivery components, i.e., Hb, O_2 saturation, and cardiac output. If oxygen extraction is standard, there is no physiologic reasoning to increase oxygen delivery by transfusion. Critical oxygen delivery at which anaerobic metabolism sets in and causes lactate production is not well defined.

The ratio of oxygen uptake to delivery is known as the oxygen extraction ratio. The standard O_2ER is around 25%. This number is essential and the body tries to compensate by increasing the cardiac output to maintain DO_2 when there is a drop in Hb [5]. It is imperative to understand critical oxygen delivery as we now have enough evidence pointing towards euvolemic management rather than increasing the hemoglobin concentration to help with oxygen delivery. We can argue that increasing the inhaled oxygen and keeping a patient on supra-normal FiO_2 can increase oxygen delivery, but as noted in the equation above, this intervention does not increase the dissolved oxygen component by much. Hence, cardiac output and hemoglobin concentration plays the primary role in oxygen delivery. Furthermore, we know from animal studies that there is a constant oxygen extraction ratio even under stress, so to secure adequate oxygen delivery or tissue perfusion, the body adjusts by ensuring enough blood flow and hence volume. DO_2 is approximately three times and 1.5 times VO_2 for the brain and heart muscle, respectively. Brain, heart, and skeletal muscle can maintain adequate blood flow in the setting of varying blood pressure within limits – the phenomenon known as 'autoregulation' [6].

Transfusion of blood products is a daily occurrence in many medical ICUs with a predefined goal to improve tissue perfusion. A decline in Hb is likely to cause a decrease in DO_2 if cardiac output remains unchanged since DO_2 = Cardiac output × CaO_2. When delivery is decreased, the body can compensate by increasing VO_2 to a certain extent. Since DO_2 incorporates Hb, cardiac output, and arterial oxygen saturation, measuring central venous O_2 ($ScvO_2$) can help determine oxygen extraction. The constant relationship between VO_2-DO_2 keeps the body functioning at an optimal level, and the 'critical level' at which this dependency is lost leads to tissue dysoxia. This critical level at which tissue dysoxia is usually present when central venous oxygen falls below 40–50% or DO_2:VO_2 is 2:1 (**Figure 1**).

If VO_2 is low or normal (<33%), the rationale to increase oxygen delivery is disputable. On the other hand, if extraction is more than 40%, this is critical, and an attempt should be to increase delivery. Simple means of measuring oxygen extraction would be to obtain central venous Oxygen concentration. Central venous oxygen concentration can be obtained simply from an indwelling central catheter in internal Jugular or subclavian vein, and this should not be mistaken for a mixed venous oxygen concentration. Mixed venous oxygen concentration (MVO_2) from an indwelling pulmonary artery catheter measures the oxygen extraction from both upper and lower extremities. However, both central and mixed venous oxygen concentrations correlate well in assessing sufficient perfusion and are often used interchangeably.

There are multiple studies available that suggest the use of this physiological parameter than an arbitrary lab-based number as a trigger for blood products is a better approach. Kocsi *et al.* performed a study on anesthetized pigs, looking at the capability of $ScvO_2$ as a parameter to reflect changes in the VO_2-DO_2 relationship.

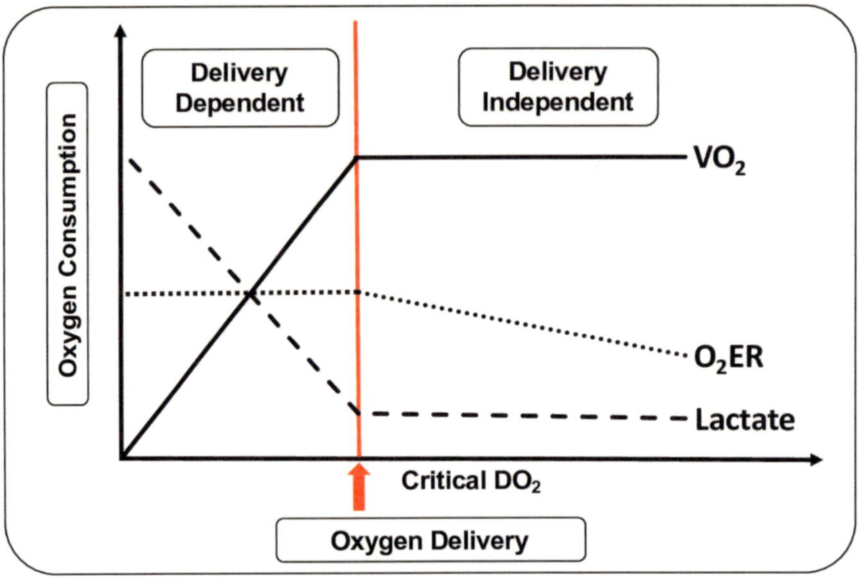

Figure 1.
Critical O_2 delivery. The body responds by extracting more oxygen in cases of decreased DO_2. However, once DO_2 drops below the critical threshold cellular metabolism becomes anaerobic with the subsequent production of lactate.

As $ScvO_2$ may reflect imbalances in delivery and consumption, this study aimed to investigate the value of $ScvO_2$ as an indicator of oxygen balance in isovolemic anemia. This study showed that $ScvO_2$ reflects changes of VO_2/DO_2 in isovolemic anemia better than Hb alone [7]. Adamczyk *et al.* studied the value of $ScvO_2$ as a trigger for transfusion in post-surgical stable patients and followed French guidelines for blood transfusion (2003). They noticed that the $ScvO_2$ increased significantly (from 57.8 to 68.5%) in the group with initial $ScvO_2$ less than 70%, whereas it was unchanged in patients with initial $ScvO_2$ greater or equal 70% (from 76.8 to 76.5%) following blood transfusion [8].

4. Newer modalities in assessing volume status

Hemodynamic monitoring has essentially become the cornerstone in guiding resuscitation for ICU patients. This serves as guide and marker for impending crisis and help determine the therapy that is best fitting. The parameters defined herein are to assist in the bedside assessment and only acts as a supplement to clinical judgment i.e., it should be considered in the context of pathophysiology, and the time point in the disease process [9]. An accurate assessment of volume status and its management remains a challenging issue with clinicians. Stroke volume is the amount of blood ejected (ml) from the left ventricle with each cardiac cycle and is the ultimate response we are trying to improve during resuscitation. Clinicians predict probable response to a fluid challenge by considering various static and dynamic parameters. The age-old tradition of using 'fluid challenge' and assessing clinical response has now been contested and multiple studies have shown increased morbidity in surgical and nonsurgical patients when not guided by a parameter. Using parameters such as cardiac index or mixed venous oxygen saturation alone do not help in guiding optimal therapy either. Gattinoni et al. showed that achieving higher levels of oxygen delivery through supranormal cardiac index or normal values of mixed venous oxygen saturation do not change mortality [10].

The static parameters previously used to assess fluid status have proven to be unreliable. Static measurements such as central venous pressure (CVP) or pulmonary artery occlusion pressure (PAOP) have low sensitivity and specificity when used to assess fluid responsiveness. Osman et al. showed that baseline CVP was similar in responders and non-responders by subjecting 96 septic mechanically ventilated patients to fluid boluses. Hence CVP measurements have a low predictive value. Also, many clinicians assume that increases in CVP with a fluid bolus indicate positive fluid responsiveness. The poor reproducibility is explained by the Frank-Starling curve, which varies among patients depending on cardiac function. These parameters are also affected by the change in respiratory system compliance and other cardiac factors. An example of this is the study done by Rivers et al. In the early goal-directed therapy arm of the trial wherein fluids administered targeted a goal CVP measurement [11]. The physiological goal of the fluid administration is to increase preload accompanied by an increased stroke volume and hence cardiac output. A notable increase in CVP with minimal change in the cardiac output or a slight change in CVP but with a notable change in cardiac output can be misleading and used alone can be confusing for clinicians. Targeting CVP, although tempting, should be combined with the after-effects, which include measurement of cardiac output in real-time.

Dynamic measurements are gaining popularity in the current era, and various modalities are available, including passive leg raise testing (PLR), stroke volume variation (SVV), and pulse pressure variation (PPV). PLR although can be used in spontaneously breathing patients, SVV and PPV come with some restrictions and best served in patients on positive pressure ventilation. A systematic review demonstrated that dynamic parameters can be reliably used in predicting fluid responsiveness. Pulse pressure variation during the Valsalva maneuver (ΔPPV) of 52% (AUC ± SD: 0.98 ± 0.03) and passive leg raising-induced change in stroke volume (ΔSV-PLR) greater than 13% (AUC ± SD: 0.96 ± 0.03) showed the highest accuracy to predict fluid responsiveness in spontaneously breathing patients [12]. Technological advances have allowed us to obtain arterial waveform analysis non-invasively or invasively in the modern-day ICU. By studying the arterial waveform, we can obtain stroke volume variation (SVV) and pulse pressure variation (PPV). A systemic review by Marik et al. demonstrated that fluid responsiveness using area under the curve (AUC) demonstrated the superiority of arterial waveform technology as compared to central venous catheters and transpulmonary dilution. Arterial waveform analysis is instrumental in the intensive care unit (ICU) and key in resuscitation for our ICU patients who already have an arterial line placed.

Dynamic parameters should be used preferentially over static parameters to predict fluid responsiveness in ICU patients. An analysis of 12 studies showed that static parameters such as right atrial pressure, pulmonary artery occlusion pressure, right ventricular end-diastolic volume were not significantly lower in responders vs. non-responders to fluid bolus [13].

4.1 Passive leg raise (PLR) testing

A relatively easy bedside test that essentially transfers 300 mL of venous blood to the right heart acts as a fluid bolus. It has significant advantages as it does not require any special equipment, easily performed at bedside and rapidly reversible with no risk of fluid overload. An important point is to directly measure cardiac output before and after the PLR test to assess if there was a notable change to cardiac output. The technique used to measure the cardiac output needs to assess short-term and transient changes to the cardiac output because the effects of PLR usually go away after one minute. This procedure can be performed by sitting the patient at 45 degrees head up in a semi-recumbent position. We lower the patient's upper body to horizontal and passively raise legs to 45 degrees. This position is sustained for about 30–60 seconds. An assessment of stroke volume (using cardiac output monitor) or pulse pressure change is conducted simultaneously.

A PLR increase of aortic blood flow higher or equal to 10% predicted fluid responsiveness with a sensitivity of 97% and a specificity of 94%. A PLR increase of pulse pressure higher or equal to 12% predicted volume responsiveness with significantly lower sensitivity (60%) and specificity (85%) [14]. A meta-analysis of 21 studies, including more than 900 patients in which cardiac output was measured by echocardiography, pulse contour analysis, bioreactance, esophageal Doppler, transpulmonary thermodilution, or pulmonary artery catheter, showed PLR-induced changes in cardiac output to have a pooled sensitivity of 0.85 and a pooled specificity of 0.91. The area under the curve was 0.95. The best threshold was a PLR-induced increase in cardiac output of greater than 10 ± 2%. For the PLR induced changes in pulse pressure, the pooled sensitivity was 0.56 (0.49–0.53), the pooled specificity was 0.83 (0.77–0.88), and the pooled area under the ROC curve was 0.77 [15]. Physiologically informed volume resuscitation with the use of the PLR-induced stroke volume change to guide management of shock is safe and demonstrated lower net fluid balance and reductions in the risk of renal and respiratory failure. In this modified intent-to-treat analysis that included 83 intervention and 41 usual care eligible patients, fluid balance at 72 hours or ICU discharge was significantly lower (−1.37 L favoring the intervention arm; 0.65 ± 2.85 L intervention arm vs. 2.02 ± 3.44 L usual care arm; P = 0.021). Fewer patients required renal replacement therapy (5.1% vs. 17.5%; P = 0.04) or mechanical ventilation (17.7% vs. 34.1%; P = 0.04) in the intervention arm compared with usual care [16].

4.2 Stroke volume variation (SVV) and pulse pressure variation (PPV)

The rationale for using SVV and PPV is that a varying left ventricular stroke volume as a response to cyclic positive pressure indicates that both ventricles are pre-load dependent as described here. SVV is derived from pulse contour analysis and the variation of the amplitude of the pulse oximeter waveform, and PPV is derived from analysis of the arterial waveform.

SVV is defined as a change in volume during the respiratory cycle and calculated as (SV max-SV min)/SV mean. This change represents dynamic variation occurring as a result of changing pleural pressure during a breath. During a positive pressure breath, intermittent increase in the intrathoracic pressure as ventilator cycles causes increased right ventricular afterload and decreased right ventricular preload. Clinically, this manifests as a decrease in left ventricular output and stroke volume after several cardiac cycles. This pressure manages to increase left ventricular preload and reduces left ventricular afterload, resulting in an acute increase in the left ventricular stroke volume harmonizing with the inspiratory phase of positive pressure ventilation. The variability in stroke volume occurring here has been well correlated with fluid responsiveness (**Figure 2**) [17, 18].

SVV has been widely studied in the operating room as it provides ideal conditions. A study by Willars et al., including high risk vascular surgeries compared SVV, PPV, CVP and Delta CVP [19]. The area under the receiver operator curve (AUROC) was 0.75 for SVV, 0.67 for PPV. The best cutoff value for SVV was 13.5%. At this level positive likelihood ratio was 2.7 and negative likelihood ratio of 0.34. They concluded that SVV was the only adequate predictor of fluid responsiveness in this cohort. SVV by itself has some limitations and needs to be used on a case by case basis.

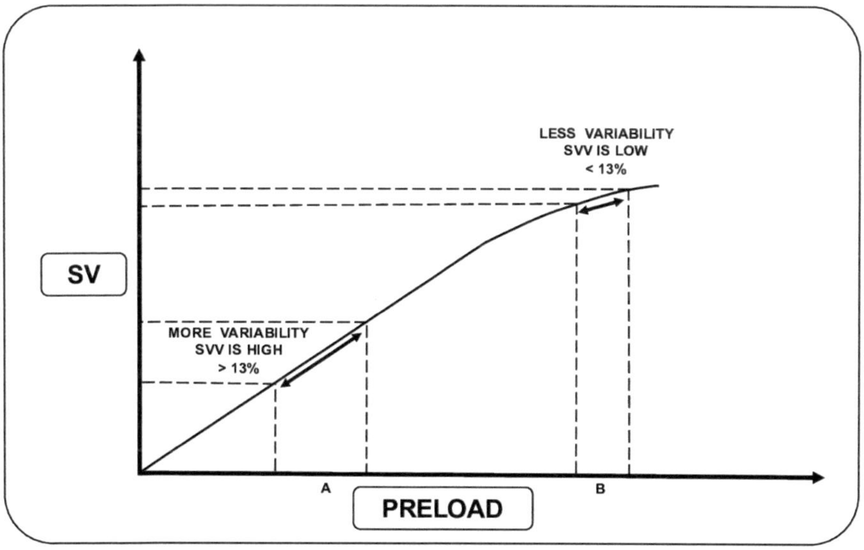

Figure 2.
SVV and Frank-Starling curve. SVV is a dynamic variable that predicts volume responsiveness and individualizes the Starling relationship. A: Volume responsive SVV > 13% B: Non-volume responsive SVV < 13%.

5. Physiologic algorithm guiding resuscitation

With various parameters now available, we suggest use of a simple physiologic algorithm to help clinicians guide therapy better for these patients. A 'volume responsive patient' can be defined using SVV, PPV, and Stroke volume index (SVI). Although there is not a true number that differentiates a true responder to non-responder, a change of 13% is thought to be appropriate. The reason being SVV is not a binary but more typically linear. Studies have defined the range to be around 10–15%. An algorithm has been proposed in detail going over the pathways that define a comprehensive approach in these patients.

In brief, dynamic changes are best observed with a significant rapid volume challenge. Colloid including blood or larger or lesser amounts of crystalloid may be used depending on the clinical scenario. In patients with SVV less than 13% you can further calculate stroke volume index (SI) as an aid to decide on the most appropriate intervention based on underlying pathophysiology. Because of the influence of heart rate on cardiac output, the preference is to use stroke index data as the measure of cardiac performance (**Figure 3**) [20].

Using this algorithm in ICU patients is quite challenging as it does not consider the rate of oxygen extraction and if volume is likely to alter it. Through advanced hemodynamic monitoring using a combination of SVV and Oxygen delivery we can further understand a patient's requirements better and an approach can be determined to patient's care. Most clinical studies were done in patients on positive pressure therapy with normal chest wall compliance, larger tidal volumes and with passive breaths. It should be considered that spontaneous breathing and arrhythmias will lead to misinterpretations of the respiratory variations in the pulse pressure/stroke volume. However, PLR has a higher sensitivity in these two scenarios and still holds value in assessment [14, 15].

The clinician must evaluate for SVV, SI and O_2 extraction on a case by case basis. Coupling these parameters assists in determination of adequate Oxygen delivery.

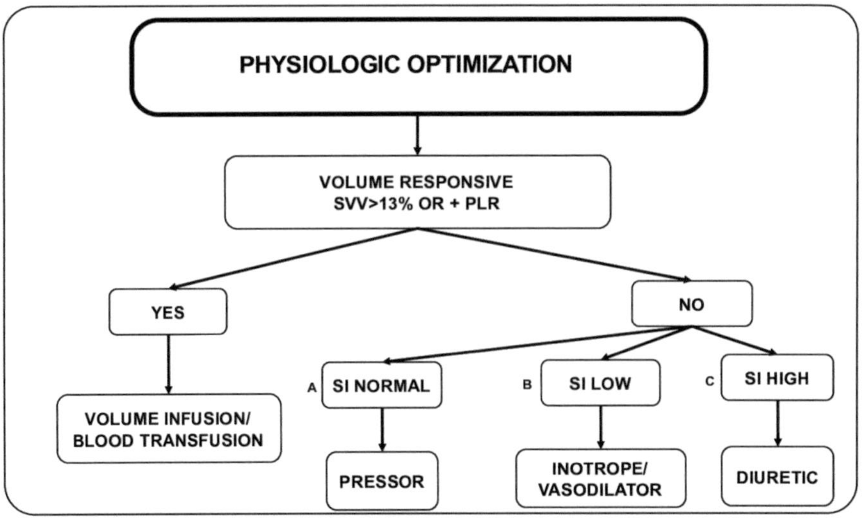

Figure 3.
Physiologic optimization. If SVV < 13% or passive leg raise test is negative, pathway A, B, C further utilizes stroke index (SI) to delineate therapy.

6. Conclusion

Physiological concepts, such as oxygen delivery and oxygen extraction, when combined with dynamic indices of fluid responsiveness, provide clinicians a more accurate assessment in predicting which cohort of patients accept additional blood or volume without resulting in lung injury. After the TRICC trial, it is a widespread practice to give blood if the Hemoglobin is less than 7 g/dL. However, if physiological parameters mentioned above are not examined, we may give blood or volume to people who do not need it and cannot accept it. By employing a simple physiologic algorithm, we can determine a specific population that will accept volume and prevents unwanted consequences of transfusion-associated lung injury and circulatory overload. This scenario prevents patients from harm and decreases various adverse outcomes, including acute lung injury, increased total body volume, higher morbidity, and mortality.

Author details

Nakul Ravikumar[1*], Geoffrey R. Sheinfeld[2] and William T. McGee[1]

1 Division of Pulmonary and Critical Care, University of Massachusetts Medical School, Springfield, MA, USA

2 Boston, MA, USA

*Address all correspondence to: nakul.ravikumar@baystatehealth.org

References

[1] Jean Louis Vincent, et al. Anemia and blood transfusion in critically ill patients. JAMA 2002;288(12):1499-1507. DOI:10.1001/jama.288.12.1499

[2] Shailaja J. Hayden, et al. Anemia in critical illness. American Journal of Respiratory and Critical Care Medicine 2012;185(10):1049-1057. DOI:10.1164/rccm.201110-1915CI

[3] A Multicenter, Randomized, Controlled Clinical Trial of Transfusion Requirements in Critical Care | NEJM. Accessed July 17, 2019. https://www.nejm.org/doi/full/10.1056/NEJM199902113400601

[4] Matthew A. Chong, et al. Should transfusion trigger thresholds differ for critical care versus perioperative patients? A meta-analysis of randomized trials. Critical Care Medicine 2018;46(2):252-263. DOI:10.1097/CCM.0000000000002873.

[5] Russell S. Roberson and Elliott Bennett-Guerrero. Impact of red blood cell transfusion on global and regional measures of oxygenation. The Mount Sinai Journal of Medicine, New York 2012;79(1):66-74. DOI:10.1002/msj.21284

[6] Christopher B. Wolff. Normal cardiac output, oxygen delivery and oxygen extraction. Advances in Experimental Medicine and Biology 2007;599:169-182. DOI:10.1007/978-0-387-71764-7_23

[7] S. Kocsi, et al. Central venous oxygen saturation is a good indicator of altered oxygen balance in isovolemic anemia. Acta Anaesthesiologica Scandinavica 2012;56(3):291-297. DOI:10.1111/j.1399-6576.2011.02622.x

[8] S. Adamczyk, et al. Contribution of central venous oxygen saturation in postoperative blood transfusion decision. Annales Francaises D'anesthesie Et De Reanimation 2009;28(6):522-530. DOI:10.1016/j.annfar.2009.03.013

[9] Michael R. Pinsky and Didier Payen. Functional hemodynamic monitoring. Critical Care 2005;9(6):566. DOI:10.1186/cc3927

[10] Luciano Gattinoni, et al. A trial of goal-oriented hemodynamic therapy in critically ill patients. New England Journal of Medicine 1995;333(16):1025-1032. DOI:10.1056/NEJM199510193331601

[11] Emanuel Rivers, et al. Early goal-directed therapy in the treatment of severe sepsis and septic shock. New England Journal of Medicine 2001;345(19):1368-1377. DOI:10.1056/NEJMoa010307

[12] Renato Carneiro de Freitas Chaves, et al. Assessment of fluid responsiveness in spontaneously breathing patients: A systematic review of literature. Annals of Intensive Care 2018;8. DOI:10.1186/s13613-018-0365-y

[13] Frédéric Michard and Jean-Louis Teboul. Predicting fluid responsiveness in ICU patients: A critical analysis of the evidence. Chest 2002;121(6):2000-2008. DOI:10.1378/chest.121.6.2000

[14] Xavier Monnet, et al. Passive leg raising predicts fluid responsiveness in the critically ill. Critical Care Medicine 2006;34(5):1402-1407. DOI:10.1097/01.CCM.0000215453.11735.06

[15] Xavier Monnet, Paul Marik, and Jean-Louis Teboul. Passive leg raising for predicting fluid responsiveness: A systematic review and meta-analysis. Intensive Care Medicine 2016;42(12):1935-1947. DOI:10.1007/s00134-015-4134-1

[16] Ivor S. Douglas, et al. Fluid response evaluation in sepsis hypotension and

shock: A randomized clinical trial. Chest 2020;158(4):1431-1445. DOI:10.1016/j.chest.2020.04.025

[17] Michael R. Pinsky. Cardiopulmonary interactions: Physiologic basis and clinical applications. Annals of the American Thoracic Society 2018;15(Suppl 1): S45–S48. DOI:10.1513/AnnalsATS.201704-339FR

[18] J. W. Biondi, et al. The effect of incremental positive end-expiratory pressure on right ventricular hemodynamics and ejection fraction. Anesthesia and Analgesia 1988;67(2):144-151.

[19] C. Willars, et al. Functional haemodynamic monitoring: The value of SVV as measured by the LiDCORapid™ in predicting fluid responsiveness in high risk vascular surgical patients. International Journal of Surgery (London, England) 2012;10(3):148-152. DOI:10.1016/j.ijsu.2012.02.003

[20] William T. McGee. A simple physiologic algorithm for managing hemodynamics using stroke volume and stroke volume variation: Physiologic optimization program. Journal of Intensive Care Medicine 2009;24(6): 352-360. DOI:10.1177/0885066609344908

Chapter 3

Four-Dimensional Flow Magnetic Resonance Imaging and Applications in Cardiology

Patrick Geeraert, Hansuk Kim, Safia Ihsan Ali, Ashifa Hudani, Shirin Aliabadi, Monisha Ghosh Srabanti, Hourieh Jamalidinan and Julio Garcia

Abstract

Blood flow through the heart and great vessels moves in three dimensions (3D) throughout time. However, the assessment of its 3D nature has been limited in the human body. Recent advances in magnetic resonance imaging (MRI) allow for the comprehensive visualization and quantification of in-vivo flow dynamics using four-dimensional (4D) flow MRI. In addition, this technique provides the opportunity to obtain advanced hemodynamic biomarkers such as vorticity, helicity, wall shear stress (WSS), pressure gradients, viscous energy loss (EL), and turbulent kinetic energy (TKE). This chapter will introduce 4D flow MRI which is currently used for blood flow visualization and advanced quantification of cardiac hemodynamic biomarkers. We will discuss its advantages relative to other in-vivo flow imaging techniques and describe its potential clinical applications in cardiology.

Keywords: Cardiac flow, 4D flow MRI, hemodynamic biomarkers, and flow quantification

1. Introduction

Imaging and quantifying various characteristics of blood flow throughout the heart is essential in modern-day cardiology. Measuring blood velocities, pressure gradients, regurgitation, stasis (and much more) is one of the most important tools physicians have for diagnosing cardiovascular pathology, stratifying severity, evaluating disease progression, and determining the most effective treatment strategies. Improving the accuracy and depth of such hemodynamic measurements is an ongoing process that continues to enhance clinical success. Two-dimensional phase contrast magnetic resonance imaging (2D PC-MRI) and Doppler echocardiography are currently the most widely used techniques for measuring cardiovascular blood flow in-vivo [1]. However, while these modalities provide immense value in clinical practice, they have their limitations. Velocity can only be measured in one direction; in Doppler echocardiography following the direction of the ultrasound beam and in 2D PC-MRI following the encoding direction assigned by the user. This can cause errors in flow measurements, depending on whether the beam/plane is placed at the exact location of interest and/or orthogonal to the direction of flow [2, 3].

These 2D measurements often rely heavily on mathematical assumptions that are not always valid [2]. For example, 2D calculations of pressure gradients are known to underestimate pressure recovery downstream of stenosis [4]. In addition, some techniques provide limited viewpoints of the thoracic cavity, such as trans-thoracic echocardiography and trans-esophageal echocardiography [1, 5]. It is also possible to acquire the in-plane velocities (X and Y directions) over time (two velocities + time) or the three plane velocities (X, Y, and Z directions) over time (three velocities + time).

Time-resolved three-dimensional (3D) phase-contrast magnetic resonance imaging (i.e. 4D flow MRI) is a novel non-invasive, non-ionizing imaging technique that provides accurate qualitative and quantitative assessment of blood velocity in all three principle directions [6, 7]. This allows for enhanced accuracy of previously measurable parameters obtained routinely by Doppler echocardiography, such as velocity and reverse flow, as well as the calculation of new parameters, such as wall shear stress (WSS), 3D pressure gradients, and turbulent kinetic energy (TKE). These parameters can be retrospectively visualized and quantified in volumes (rather than cross sections), over the course of a cardiac cycle, and in unlimited viewpoints. Authors can refer to 2D, 3D, 4D, 5D or 7D flow depending on the acquisition scheme used to encode the velocity directions over time. Thus, it is important to understand how the velocity acquisition is defined. In this chapter, 4D flow MRI measures 3 velocity encoded directions in a stack of planes along the cardiac cycle. As such, the ongoing development of 4D flow MRI provides great promise in improving the clinical management of cardiovascular disease.

2. Data acquisition and pre-processing

2.1 Safety and preparation

There are safety measures and recommendations that should be considered for subjects undergoing 4D flow MRI [6, 8]. Patients can fill out pre-imaging safety questionnaires that consider items that can cause a hazard or interfere with the MRI examination. The safety information may become frequently updated because of continuous and rapid changes in the MRI technology. Before starting any cardiac MRI study, compatible electrocardiogram (ECG) leads should be placed on the subject's chest properly. The accurate synchronization of data acquisition with phases of the cardiac cycle, including different stages of contraction and relaxation, is one of the essential requirements for efficacious cardiac MRI exam. This technique is called ECG-gating. A phased-array receiver coil is required to capture the electromagnetic signals needed to create an image. Thus, it is important that receiver coils are positioned appropriately to cover the regions of interest. The use of contrast agents is optional in 4D flow MRI, but they can help increase the signal-to-noise ratio, and therefore improve the image quality.

2.2 Data acquisition

In 4D flow MRI, acquisition parameters (ex: spatial resolution, temporal resolution, field of view, etc.) are optimized and programmed into the scan protocol to provide the best possible imaging accuracy. Setting major scan parameters is primarily a fine balance between adequate temporal/spatial resolution and minimized scan time. Imaging accuracy can also be affected by artifacts (i.e., distortions in the image that are not present in reality), of which velocity encoding (Venc) is an important contributor. Velocity encoding is a user-defined parameter that sets the upper and lower limits of blood velocity which can be appropriately imaged for within the scan protocol. If the patient's blood velocities fall above the Venc limit, aliasing will corrupt the image with artifacts in those locations. However, if the Venc range is set too high, the image can be populated with noise [9]. Furthermore, artifacts can arise from movement due to cardiac motion and breathing. As shown in **Figure 1**, ECG-gating compensates for cardiac motion to determine when the heart is most still, at which point images are acquired [10]. Simultaneously, diaphragmatic navigator gating compensates for breathing artifacts by similarly tracking movement of the patient's diaphragm and acquiring images at the point of least movement [11]. This allows the patient to breath freely during the scan without creating breathing-related artifacts. These approaches are applied to provide the clearest possible images of 3D global and local blood flow characteristics. Four types of images are produced after acquisition: one magnitude image (shows anatomical structures) and three phase images (shows blood velocity along the Venc directions, often x, y, and z axes).

The most important limiting factor for adding 4D flow to a routine clinical cardiac MRI exam is long scan times associated with multidimensional imaging over the entire cardiac cycle. In 1990s, the total scan time was roughly 40–80 minutes which made difficult its routine application in clinical settings. In recent years, scan times have been further reduced due to ongoing progress in advanced imaging techniques. Nowadays, different reconstruction techniques (parallel imaging, radial and Cartesian sampling, compressed sensing, etc.), in addition to enhanced computing power, have reduced the scan time to 3–10 minutes [12–14]. The latter is facilitating is penetration as a diagnostic tool. Finally, all acquired data are saved in the Digital Imaging and Communications in Medicine standard format in the MRI system database.

2.3 Pre-processing and correction

Due to a range of errors, image quality can be damaged by various factors including noise, eddy current effects, concomitant gradient field effects (Maxell terms), velocity aliasing, and gradient field nonlinearity. Data pre-processing is applied to rectify these potential errors using several correction strategies to make 4D flow MRI a reliable source of 3D flow visualization and quantification, as illustrated in

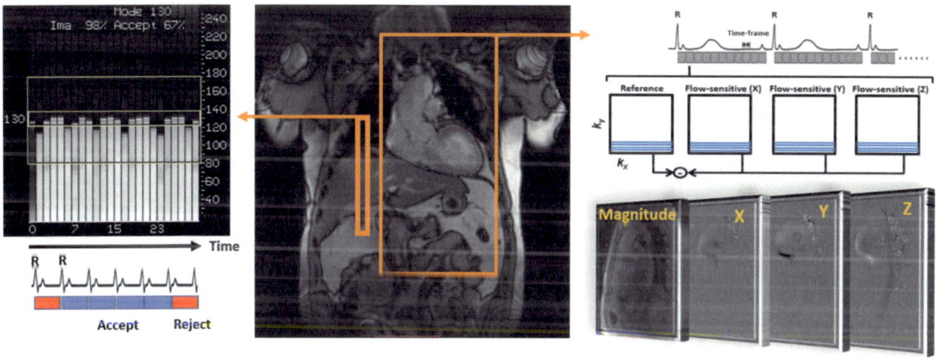

Figure 1.
Data acquisition for 4D flow MRI. *Data acquisition covering the whole heart (large orange rectangle) is acquired using electrocardiogram-gating and respiratory control (small orange rectangle). Three-dimensional velocity-encoding (right side) is used to obtain velocity-sensitive phase images which are subtracted from reference images to calculate blood flow velocities along all three spatial dimensions (X, Y, Z) and averaged magnitude visualizing anatomy over the cardiac cycle.*

Figure 2 [15]. Three-dimensional phase-contrast magnetic resonance angiography (3D PC-MRA) can be obtained at this stage based on acquired data from 4D flow MRI by several presented strategies without the need for further MRI acquisition. A 3D PC-MRA can display complex vascular structures and geometries of interest without requiring a contrast agent. It is vastly helpful in some situations, such as patients with contrast agent contraindication. In addition, 3D PC-MRA allows the user to retrospectively isolate specific volumes of interest for analysis.

However, 4D flow MRI images present difficulties for segmentation algorithms due to extensive variability in cardiovascular structure, geometric intricacy, low resolution, high background noise and motion artifacts. For that reason, manual segmentation remains a widely used method, but manual segmentation takes a long time to perform and is prone to observer variability [16]. There are some established semi-automatic segmentation methods [17–19] which are faster than manual segmentation, nevertheless they are still operator dependent. Recent machine learning and artificial intelligence strategies have shown great ability to solve 4D flow MRI segmentation problems [20]. Machine learning algorithms are powerful techniques that train a machine (i.e. computer) to perform a specific task. Convolutional neural networks (CNNs), which is one of the deep learning techniques, forms the foundation of image segmentation. U-Net is a CNN that was developed specifically for medical image segmentation [21]. Several recent studies that focused on advancing imaging segmentation techniques. Berhane et al. developed a CNN model to segment the aorta from 4D flow MRI images [22]. The performance in this study was compared with manual segmentations, and they reported good agreement across flow and diameter along the aorta. Segmentation speed was <1 second per case, while manual segmentation required at least 360 seconds. Bratt et al. suggested a network based on the U-net and residual modules [23]. They reported similar segmentation performance, < 0.6 seconds per case, however the manual segmentation required 238 seconds per case. Wu et al. developed a combinatorial network for segmenting the left ventricle (LV) [24]. They claimed that combining networks can increase the accuracy of the segmentation.

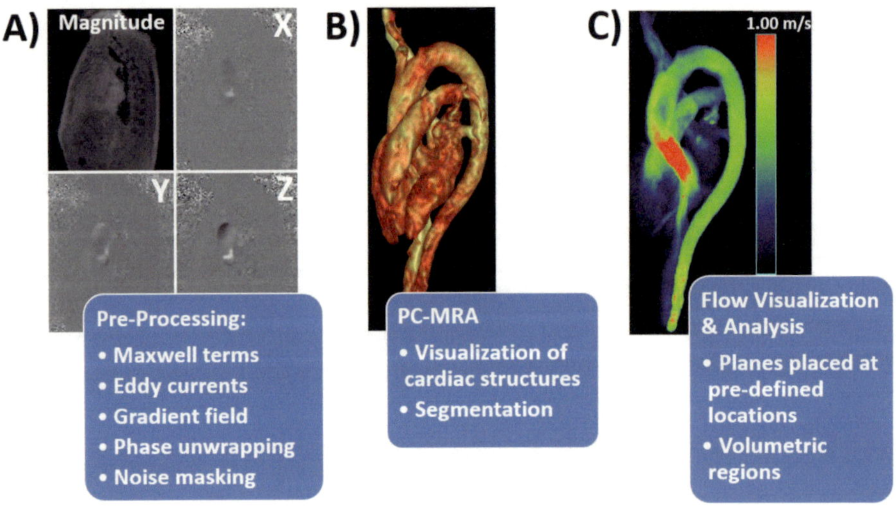

Figure 2.
Data pre-processing. *Acquired 4D flow images are pre-processed applying multiple corrections (panel A). A phase-contrast angiogram (PC-MRA) is calculated to visualize the cardiac structures and can be used for individual segmentation of vessels, panel B. the generated PC-MRA and/or segmentation can be used to mask the velocity field for appropriate visualization and analysis using planes or volumetric regions of interest, panel C. In this example a 58-year-old control volunteer is presented.*

2.4 Flow visualization and quantification

Before flow visualization and quantification, general data quality control, including visual inspection and quantitative quality control, is recommended to ensure internal data is consistent and accurate. Dedicated visualization and quantification software can be used for obtaining streamlines, pathlines, volume rendering, and/or maximum intensity projection (MIP). Streamlines represent a blood particle's instantaneous path tangential to the velocity vector at a particular point of time, such as at peak systole. Pathlines represent a blood particle's path over time (i.e. trajectory). Volume rendering allows us to represent voxel-by-voxel values in a dynamic spatial manner, given access to the entire field of view. A MIP is a single-plane image representing the maximum values through a given direction of the volume, similar to an X-ray image. Taken together, these flow visualization techniques reveal a wealth of information about blood flow abnormalities and cardiovascular disease progression. One of the most significant advantages 4D flow MRI is the ability to retrospectively quantify cardiac flow parameters within specified regions of interest. Flow can be analyzed within a specific volume that has been isolated via segmentation, or via a 2D cross-sectional analysis plane placed within a volume of interest. These analysis planes can be flexibly placed at any anatomical location to quantify general and advanced blood flow parameters. Some of these visualization and quantification methods will be illustrated in the following section.

3. Applications in cardiology

3.1 Congenital heart disease

Eight out of every 1000 infants are born with congenital heart disease (CHD), which encompasses all structural heart defects present at birth, including the great vessels and cardiac valves [25]. Individuals with CHD may develop many cardiac complications, even after surgical correction of the abnormality, including valve insufficiency, arrhythmias, and heart failure [25]. Tetralogy of Fallot is one of the most common forms of cyanotic ("blue baby") CHD, accounting for about 10% of all CHD [26]. Tetralogy of Fallot involves a combination of four defects including right ventricular hypertrophy, aortic override, pulmonary stenosis, and a ventricular septal defect. These patients undergo multiple repeat surgeries and procedures over their lifetime but the hemodynamic factors contributing to the optimal quality of life and outcomes are understudied and poorly understood. Congenital heart disease can benefit from 4D flow MRI via the calculation of advanced hemodynamic parameters, such as WSS, TKE, 3D pressure mapping, and energy loss [5]. For demonstration, we primarily focus on pressure mapping, but **Table 1** provides an overview of 4D flow MRI hemodynamics in CHD.

Fluid pressure measured within the cardiovascular system is widely used in CHD diagnosis, such as coarctation of the aorta, pulmonary hypertension, or atrio-ventricular septal defects. The pressure difference across structural abnormalities (ex: stenosis or ventricular septal defect) or within the LV can reveal much about the severity of the disease. Pressure mapping, based on the measured 3D blood flow velocity field, can be calculated by solving the Navier–Stokes equation, which describes the time-varying flow of a viscous, incompressible Newtonian fluid [33, 34]. This method allows for the estimation of temporally and spatially distributed pressure gradients and across a large vessel segment or cardiac chamber, **Figure 3**. Overall, 3D pressure mapping allows us to gain a better understanding of

Region	CHD type	Conventional flow parameters	Advanced flow parameters	References
Venous return	Fontan/single Ventricle	Collateral flow volume, peak velocity, valvular flow volume	KE, EL, flow connectivity distribution	[5, 8, 27, 28]
	Tetralogy of Fallot	Right heart (RA, RV, PA) systolic peak velocity, net flow, retrograde flow	WSS, vorticity, KE, EL, TKE, pressure drop	[5, 7, 8, 27–29]
Heart wall	Atrial Ventricular Septal Defects	Flow volume, shunt flow volume, shunt ratio	Vorticity, KE, EL, WSS	[5, 8]
Aortic valve	Bicuspid Aortic Valve	Net flow, regurgitation volume, peak velocity	WSS, turbulence, EL, pressure mapping, helicity, flow eccentricity, PWV	[5, 7, 8, 27, 28, 30]
	Marfan Syndrome	AV Peak flow, AV flow volumes	WSS, helical flow, PWV, pressure mapping	[8, 30, 31]
Outflow tracts	Dextro Transposition of the Great Arteries	Net flow volumes, flow ratio, peak velocity	WSS, helical flow	[5, 8]
	Aortic Coarctation	Collateral flow volume, peak velocity across stenosis	Pressure mapping, EL, WSS, helical flow	[5, 8, 27, 30, 32]

CHD: congenital heart disease; AV: aortic valve; RA: right atrium; RV: right ventricle; PA: pulmonary artery; EL: viscous energy loss; KE: kinetic energy; PWV: pulse wave velocity; TKE: turbulent kinetic energy; WSS: wall shear stress.

Table 1.
Overview of research on conventional and advanced 4D flow–derived hemodynamic parameters in CHD.

the basic determinants of time-varying flow in healthy and diseased hearts, which has the potential to improve our methods for diagnosing, treating, and surgically correcting CHD [35, 36].

3.2 Mitral regurgitation

Approximately 10% of the general population will develop mitral regurgitation (MR) throughout their lifetime [37]. Mitral regurgitation is defined as systolic retrograde flow from the LV into the left atrium (LA). The main causes are classified as primary (structural or degenerative abnormality of mitral valve apparatus) or secondary (a disease of the LV interferes with function and integrity of mitral valve apparatus) [38]. This disorder generally progresses insidiously, as the heart compensates for increasing regurgitant volume via LA enlargement that partially relieves LV overload. Long-standing regurgitation yields poor outcomes as it may lead to progressively more severe regurgitation, LV failure, pulmonary hypertension, atrial fibrillation (AF), stroke, and death [39]. If MR is severe, surgery is the recommended treatment to prevent heart failure [1]. Thus, early detection and assessment is crucial, especially in elderly patients who are often ineligible for surgical intervention.

Doppler echocardiography is the most widely used tool for the assessment of MR. The commonly considered parameters when classifying MR severity are jet area, width of vena contracta, effective regurgitant orifice area, regurgitant volume,

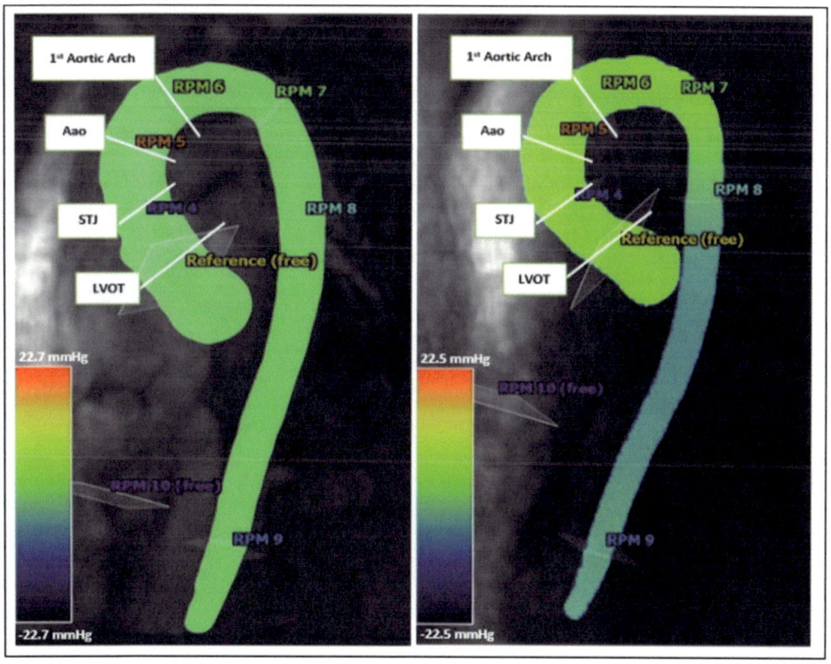

Figure 3.
Pressure mapping. *Volume rendering maps of a control (left) and a patient with repaired tetralogy of Fallot (right). Several analysis planes, including location reference. Reference plane for pressure is in yellow. RPM indicates analysis plane. LVOT: Left ventricular outflow tract; STJ: Sinotubular junction; AAo: Ascending aorta.*

and regurgitant fraction. However, the accuracy of standard approaches used to quantify these parameters can be influenced by the mechanism of regurgitation, direction of the jet, jet momentum, LV loading condition, LA size, and the patient's blood pressure [40, 41]. It is also important to keep in mind that these standard-of-care diagnostic tools do not permit a comprehensive in-vivo assessment of 3D blood flow which is critical to the study of complex 3D hemodynamics surrounding MR. Cardiac MRI has recently been reported as a more accurate tool for quantification of MR flow characteristics and severity grading [42, 43]. Advanced measures of vortex formation, helical flow patterns, EL, pressure mapping, and WSS have shown usefulness for assessment of valve-related disease using 4D flow MRI [44]. The shape of vortex cores have been shown to closely resemble the valve shape, while the vortex's orientation is related to the LV inflow direction [45]. In demonstrating how to extract vortex information from 4D flow MRI, Krauter et al. showed that vortex shape, vorticity and kinetic energy (KE) strongly correlate with transmitral peak velocities [46]. Helical grade was also associated with systolic anterior motion of the mitral valve. Lastly, as shown in **Figure 4**, MR disturbs organized flow, resulting in a reduced contribution of left pulmonary veins to the vortical flow, potentially leading to less efficient ventricular filling and stasis [47].

It is important to note, however, that flow-based biomarkers still require further exploration before they can be reliably applied to daily clinical practice. The construction of 'atlases' that depict physiologically normal blood flow patterns through the LA shows great promise in helping identify the clinical utility of certain hemodynamic parameters and personalizing diagnoses. Normally, MR treatment is primarily based on chamber dimensions and qualitative regurgitation severity grading [48], but it is well recognized that these measures may be insufficient to guide treatment strategies and require multi-modality integration. Four-dimensional flow

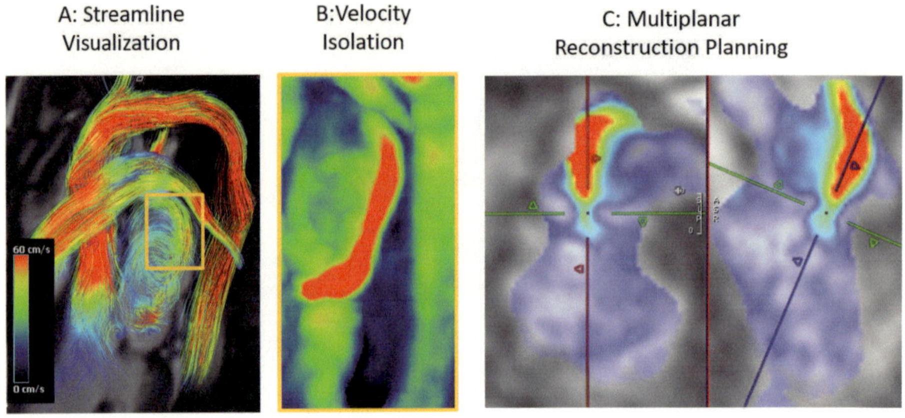

Figure 4.
Mitral regurgitation. *The presented case has mitral thickening with evidence of cleft in the anterior mitral leaflet. Panel A shows a whole heart streamline visualization used to locate mitral regurgitation (orange box). Panel B shows the velocity volume rendering of the velocity field. The mitral regurgitant jet was highly eccentric into the atrial wall. Panel C illustrates the multiplanar reconstruction of the velocity field which allows multiple views of the regurgitant jet. Evaluation led to a mild regurgitation grading with a regurgitant fraction of 15%.*

MRI has provided the ability to construct time-averaged 3D hemodynamic maps (i.e. atlases) from healthy subjects, which can then be used as a reference when evaluating a patient's MR severity. For example, Goffic et al. recently developed strategies for the generation of KE and helicity atlases [49]. Their preliminary data suggest that these atlases may provide insight into hemodynamic influence on LV dysfunction progression and, thus, could have implications on personalized assessment of MR. Such atlases can be created for any hemodynamic parameter of interest and will be further elaborated on in the aortic diseases section of this chapter.

3.3 Atrial fibrillation

Atrial fibrillation is an abnormally fast heart rhythm with uncoordinated atrial activation and ineffective atrial contraction [50, 51]. Multiple simultaneous electrical signals firing within the atria lead to irregular ECG patterns and atrial activity, loss of coordinated atrial contractions, and inadequate ventricular filling. It is classified according to the duration of episodes. At an early stage, an episode of AF terminates within 7 days of onset and sinus rhythm is restored (paroxysmal AF). However, as severity progresses, the AF episode may last beyond one week (persistent AF) or does not terminate (permanent AF). The most common complication of AF is thromboembolic events such as stroke [50, 51]. Reduced LA function increases the risk of blood stasis and clot formation in the LA, especially the left atrial appendage (LAA) which is a small extension of the LA. The CHA_2DS_2-VASc score (accounts for patient history of: Congestive heart failure, Hypertension, Age > 75 years, Diabetes mellitus, Stroke, Vascular disease, Age between 65 and 74 years, and Sex) is currently recommended for stroke risk stratification for AF patients, based on clinical factors such as age, gender, and disease history [50–52]. This risk score is used to recommend the use of anticoagulants and further therapy. However, the score does not contain other individual physiologic factors, so prediction power is limited.

There have been efforts to improve diagnosis and evaluation of disease and risk-assessment of AF through analysis based on 4D flow measurements (**Table 2**). Although some contradictory reports exist, most of the recent studies characterizing AF blood flow with relatively large cohorts agree that there is a significant decrease of LA flow velocity in both persistent and paroxysmal AF patients

Study	Cohorts (n)	LA Flow parameters	Main Findings
Fluckiger et al. (2013) [53]	PAF (6) Persistent AF (4) Controls (19)	Mean velocity	Mean velocity ↓ in persistent AF cohort
Markl et al. (2016) [54]	AF-sinus (42) AF-afib (39) Young controls (10) Controls (20)	Peak velocity, time-to-peak velocity, stasis	1. Peak velocity ↓ in AF-afib 2. Stasis ↑ in AF-sinus and AF-afib
Lee et al. (2016) [55]	AF (40) Young controls (24) Controls (20)	Velocity (mean, median, and peak)	1. Velocity ↓ (mean and median showed most significant difference) 2. CHA_2DS_2-VASc score inversely correlated with mean, median, and peak velocity
Markl et al. (2016) [56]	AF-sinus (30) AF-afib (30) Controls (15)	Velocity (mean and peak), Stasis (in LA and LAA)	1. Individual variability of flow patterns in AF patients, despite the same CHA_2DS_2-VASc score 2. CHA_2DS_2-VASc correlated positively with stasis, but negatively with velocity
Garcia et al. (2020) [57]	PAF (45) Controls (15)	LA velocity (mean, median, and peak), pulmonary vein peak velocity, stasis, vortex size	1. Mean and median LA velocity ↓, pulmonary vein peak velocity ↓ 2. Stasis ↑ 3. Vortex size ↑ and correlated with CHA_2DS_2-VASc
Kim et al. (2020) [58]	PAF (28) Controls (10)	Peak velocity, delayed ejection, residual volume, regurgitation	1. Residual volume ↓ 2. Delayed ejection ↑
Demirkiran et al. (2021) [59]	PAF (10) Controls (5)	Velocity (mean and peak), stasis (in LA and LAA), KE (mean and peak)	1. Mean/peak velocities ↓ 2. LA and LAA stasis ↑ 3. Mean and peak KE ↓
Spartera et al. (2021) [60]	AF- afib (22) AF-sinus (64)	Velocity (mean and peak), stasis, vorticity, vortex volume	Peak velocity and vorticity are reproducible, stable, and exhibit similar interval-scan variability between cohorts

LA: left atrium; LAA: left atrial appendage; PAF: paroxysmal atrial fibrillation; AF: atrial fibrillation; AF-sinus: previous history of AF, but in sinus rhythm at time of imaging; AF-afib: in AF at time of imaging; KE: kinetic energy; CHA_2DS_2-VASc: stroke risk stratification system that accounts for patient history of congestive heart failure, hypertension, age > 75 years, diabetes mellitus, stroke, vascular disease, age between 64 and 75 years, and sex.

Table 2.
Summary of 4D flow studies in AF.

[54–57, 59]. Most notably, the increase of CHA_2DS_2-VASc score has been associated with reduced mean LA velocity [55, 56], which suggests 4D flow measurement may be able to improve risk assessment. Kinetic energy, which is proportional to the mean square of velocity, was also found to be markedly lower in AF patients than in controls [59]. Left atrial flow stasis is defined by Markl et al. [54] as the relative number of time frames, for each voxel, with flow velocity less than 0.1 m/s. Flow stasis has been confirmed by several studies to be elevated in AF patients [6, 8–10]. An example of a MIP for flow stasis is displayed in **Figure 5**. Also, flow patterns through the pulmonary vein into the LA have been studied [57].

In AF patients, fragmentation of LA inflow and vortex formation in the LA was characterized, see **Figure 6**. This study demonstrated that vortex size increased in paroxysmal AF and was associated with a greater risk score. Similar findings of decreased velocity and increased stasis have also been found in the LAA specifically [56, 59]. However, the limited spatial resolution of 4D flow MRI may not satisfy the minimum number of voxels needed to segment the LAA accurately and reliably in a certain number of cases [6]. In a recent study, reliability and reproducibility of 4D flow parameters in AF patients were reported [60]. Left atrial peak velocity

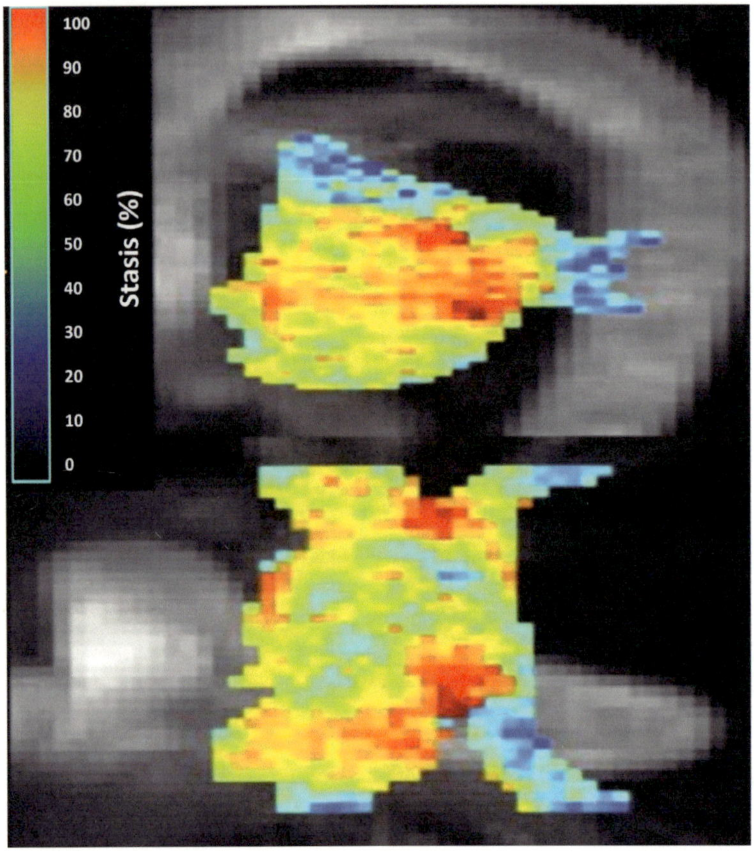

Figure 5.
Left atrial stasis maps. *Maximum intensity projections of stasis using a sagittal-view (top) and a top-view (bottom).*

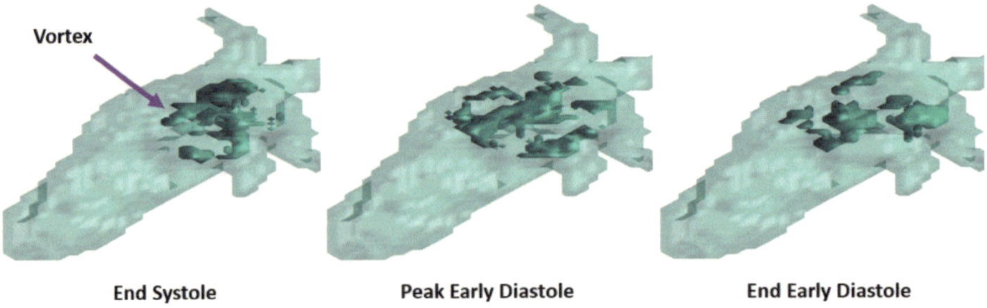

Figure 6.
Atrial vortex dynamics in atrial fibrillation.

and vorticity were found to be more reproducible and independent of physiological factors than mean velocity, vortex volume and stasis.

3.4 Bicuspid aortic valve disease and Aortopathy

Bicuspid aortic valve (BAV) disease is the most common congenital valve disease, affecting 0.5–1.4% of the general population [61–63]. While an aortic valve normally contains three functional leaflets (i.e. tricuspid), a BAV contains only two. There are different sub-types of BAV: valves that developed with only two leaflets (Type 0) and valves that developed with three leaflets containing a fusion between any adjacent pair (Type 1) [64]. Type 1 phenotypes are further subdivided depending on what leaflet pair is fused. Despite being a valvular malformation, BAV disease is closely associated with aortic dilation (BAV aortopathy) that increases patients' risk of aortic aneurysms and dissections [65]. Traditionally, aortic diameters and growth rates have been used to stratify BAV patients at risk for aortic dissection, but these measures alone have been shown to possess limited prognostic value [66]. As well, uncertainties still exist regarding the exact pathophysiology of BAV aortopathy and the most effective timing of surgical intervention [67]. Thus, it is important to study new biomarkers that may enhance our understanding of BAV disease progression. Four-dimensional flow MRI has allowed the study of several new and promising biomarkers, such as abnormal flow patterns, WSS, and energy loss.

Eccentric flow jets and helicity are two characteristics of abnormal flow patterns that have shown strong connections with aortic dilation in BAV patients. A tricuspid aortic valve typically produces a centered systolic jet and bulk flow that is parallel to the ascending aorta, while BAVs tend to produce off-centered systolic jets (eccentric flow jets) that lead to circumferential flow and vortices (helicity). Each BAV phenotype has been shown to produce its own general pattern of jet eccentricity and helicity, and the direction and orientation of these abnormal flow patterns has been associated with patterns of aortic dilation [68–71]. For example, patients with right–left coronary leaflet fusion (Type 1 RL) are more likely to produce a flow jet aimed to the right-anterior wall that associates with dilation focused at the mid-ascending aorta, while right-non coronary leaflet fusion patients (Type 1 RN) tend to produce right-posterior flow jets that associate with diffuse dilation extending to the aortic root and/or arch as well, **Figure** 7 [69, 70, 72]. Furthermore, Bissel et al. showed that BAV patients with normal flow jets and non-helical flow patterns tended to have similar aortic diameters to healthy volunteers [73]. While more longitudinal studies are needed to confirm causation, these studies seem to collectively suggest that abnormal flow patterns are connected to aortic dilation in BAV patients.

Wall shear stress, a measure of force exerted on the vessel wall by flowing blood, has consistently shown to be elevated in the ascending aorta of BAV patients [69, 74, 75]. The abnormal flow patterns created by a BAV are likely responsible for these increased WSS forces, and WSS itself has also been associated with regions of aortic dilation (**Figure** 7) [68, 69, 71, 73]. Seminal studies conducted by Bollache et al. and Guzzardi et al. demonstrated a possible physiologic mechanism behind WSS-associated aortic dilation, showing that elevated WSS levels may trigger maladaptive metalloproteinase activity which leads to medial elastin fiber degeneration and overall weaker connective tissue in the aortic wall [76]. Thus, it is thought that elevated WSS, driven by abnormal flow patterns, may be a direct mediator of aortopathy in BAV patients. This ability of 4D flow MRI to visualize flow patterns and quantify WSS may provide future clinical utility in the risk-stratification of BAV patients and identifying appropriate timing for aortic surgery.

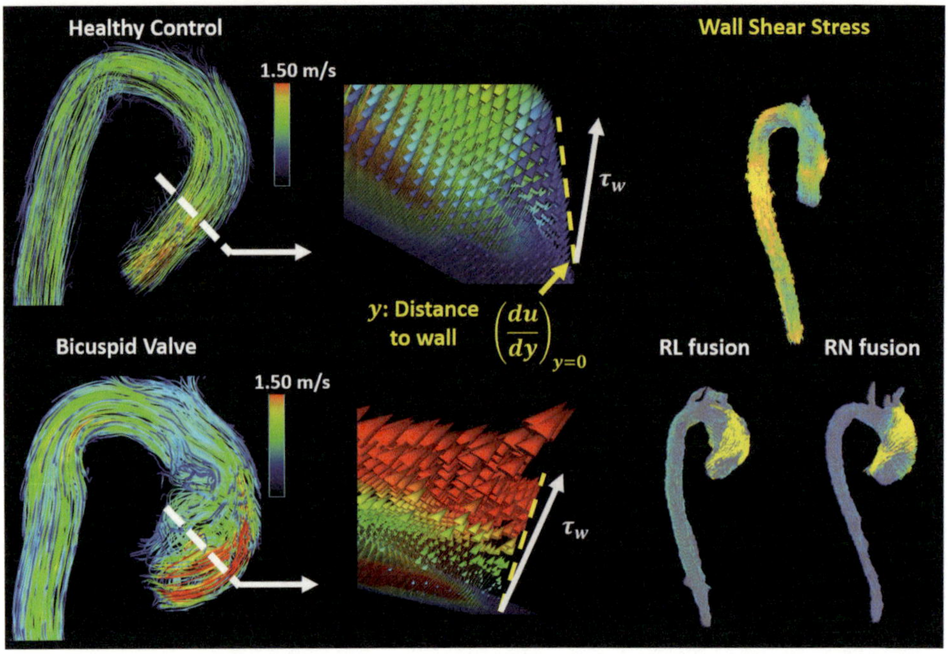

Figure 7.
Flow patterns and wall shear stress in a control and bicuspid valve phenotype. *The white dashed lines represent the location where the sample velocity profile was obtained. These flow profiles can estimate the shear stress rate, blood flow spatial deformation. Since no flow occurs through the vessel wall, the speed of the blood flow at the vessel boundary is zero. The near wall region is the boundary layer where the wall shear stress (WSS) forces occur. The WSS expresses the force per unit area exerted in the fluid direction on the local vessel tangent (τ_w). Yellow dashed lines illustrate the flow profile slope near the wall. On the top, a healthy control illustrates normal flow in the proximal ascending aorta. On the bottom, samples of right–left (RL) fusion and right-non coronary (RN) fusion illustrate abnormal flow.*

3.5 Aortic stenosis

Aortic stenosis (AS) refers to the narrowing of the aortic valve opening, which restricts blood flow from the LV to the aorta. It is the most prevalent valvular disease in developed countries, affecting 2.4% of those >75 years of age [77]. It is commonly a result of BAV disease, chronic calcification, or rheumatic fever (in developing countries). Aortic stenosis often leads to complications such as LV dysfunction, heart failure, and aortic dilation, aneurysm and dissection [78]. Surgical repair or valve replacement are the only known definitive treatments and accurate diagnosis and staging are critical for surgical decision-making [79]. The most widely used parameters in assessing valve function include transvalvular pressure gradient, peak velocity, and valve area. However, up to 40% of AS patients present with discordant findings (ex: abnormally small valve area, but normal pressure gradient) that require additional imaging, and the heterogenous nature of onset of secondary complications (ex: different dilation patterns, different rates of progression, etc.) is not well understood [80]. Four-dimensional flow MRI research continues to enhance our understanding these issues through the novel measurement of 3D peak velocity, 3D pressure gradients, and fluid energy losses.

Peak velocity and pressure gradients across the aortic valve are key components in AS severity grading [79]. However, as previously mentioned, echocardiography and 2D PC-MRI measurements often underestimate peak velocity, due to inaccurate

2D analysis plane placement, and overestimate pressure gradients, due to exclusion of downstream pressure recovery in calculations. The 3D visualization of these parameters using 4D flow MRI allows for more accurate identification of true peak velocity and a comprehensive calculation of pressure gradients that accounts for pressure recovery in the thoracic aorta. Due to this, 4D flow MRI may be a more accurate imaging modality for AS severity grading and surgical decision-making. Although, it should be noted that 3D pressure gradient calculations assume laminar flow, which may lead to inaccuracies when measuring severely stenotic flow where turbulence exists [33].

Fluid energy loss is an advanced hemodynamic parameter that provides information regarding LV workload. There are two types of mutually exclusive fluid energy loss measurements: viscous energy loss (EL; energy lost due to friction between adjacent fluid layers with different velocities) and TKE (energy lost to turbulence) [81]. Both measurements reflect LV output lost to abnormal flow patterns and, ultimately, a greater cardiac afterload. Larger fluid energy losses place greater workloads on the LV, which can lead to LV dysfunction and heart failure. Several studies have shown the presence of significantly elevated fluid energy losses in AS patients and explored the role of TKE in improved characterization of AS severity [82, 83]. Specifically, Binter et al. found TKE to be greater in AS patients compared to controls and demonstrated TKE to be influenced by aortic and valvular morphology [83]. Taken together, this body of research suggests that fluid energy loss may provide novel AS severity measurements that are complimentary to traditional evaluation techniques. Lastly, it should be noted that jet eccentricity, helicity, and WSS measurements may serve the same purpose in the study of AS as they do in the study of BAV disease. These parameters have shown close associations with aortic dilation, a common complication in AS patients [68, 69, 73, 84, 85]. Most studies exploring these associations use BAV patient cohorts, since AS is a common finding in BAV disease.

3.6 Aortic Coarctation

Aortic coarctation (CoA) refers to a congenital narrowing of the thoracic aortic lumen, most often in the arch or descending portion, and accounts for approximately 6–8% of all congenital heart defects [86]. It often accompanies other congenital malformations, such as BAV (60%), aortic arch hypoplasia (18%), ventricular septal defect (13%), and sub AS (6%) [87]. Similar to AS, CoA causes an upstream increase in pressure, which may lead to LV dysfunction, aortic aneurysm and dissection, upper body hypertension, and stroke. Current diagnostics include computed tomographic angiography to image aortic structure and echocardiography to measure peak velocities and pressure gradients across the stenotic region. Due to the limitations of echocardiography, 4D flow MRI may provide added utility in characterizing CoA via 3D measurement of peak velocity profiles and pressure gradients.

The application of 4D flow MRI in generating 3D peak velocity profiles and pressure gradients serves the same benefits as previously mentioned in other cardiovascular disease. That is, it allows the visualization of temporally and spatially resolved flow patterns so that analysis planes may be placed in the most applicable locations and pressure recovery can be accounted for when measuring pressure drop across the CoA (**Figure 8**). Several studies to date have found benefit in the use of 4D flow MRI for characterizing CoA flow patterns and evaluating post-repair hemodynamics in CoA patients [88–90].

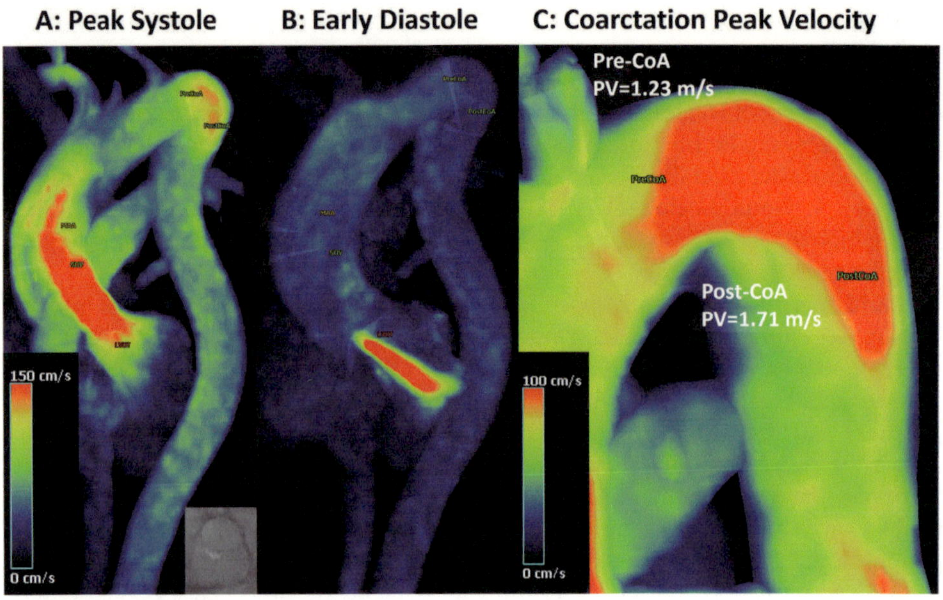

Figure 8.
Evaluation of a coarctation of the aorta. *Flow is visualized at peak systole (A) and early diastole (B). (C) Peak flow velocity (PV) across coarctation. This patient (male, 53 years old) has a type 1 LR BAV with severe aortic insufficiency (regurgitant fraction 56%), mild stenosis, coarctation measuring 21 mm, and moderate dilatation of the proximal ascending aorta (46 mm). LR: Left – Right coronary leaflet fusion; BAV: Bicuspid aortic valve.*

4. Conclusion

In conclusion, 4D flow MRI is a powerful technique which can be used for calculating important clinical parameters. This chapter intended to introduce and summarize the usefulness of 4D flow for assessing cardiovascular diseases. Thanks to recent technical advances, 4D flow MRI has increased its use in cardiac MRI sites worldwide and it is in a ready-to-go state-of-art stage for clinical practicality.

Acknowledgements

Authors were supported by The University of Calgary, URGC SEM #1054341 and JG start-up funding. Research unrestricted funding was also provided by The Libin Cardiovascular Institute and Siemens Healthineers. HK, SIA, and SA received scholarship support from the Biomedical Engineering graduate program. We acknowledge the support of the Natural Science and Engineering Research Council of Canada/Conseil de recherche en science naturelles et en génie du Canada, RGPIN-2020-04549 and DGECR-2020-00204.

Conflict of interest

Authors have no conflict of interest to declare in the context of this chapter.

Author details

Patrick Geeraert, Hansuk Kim, Safia Ihsan Ali, Ashifa Hudani, Shirin Aliabadi, Monisha Ghosh Srabanti, Hourieh Jamalidinan and Julio Garcia*
University of Calgary, Calgary, Canada

*Address all correspondence to: julio.garciaflores@ucalgary.ca

References

[1] Otto CM, Nishimura RA, Bonow RO, Carabello BA, Erwin JP, Gentile F, et al. 2020 ACC/AHA Guideline for the Management of Patients With Valvular Heart Disease: Executive Summary. J Am Coll Cardiol. 2021;77(4):450-500.

[2] Fyrenius A, Wigstrom L, Bolger AF, Ebbers T, Ohman KP, Karlsson M, et al. Pitfalls in Doppler evaluation of diastolic function: Insights from 3-dimensional magnetic resonance imaging. J Am Soc Echocardiogr. 1999;12(10):817-826.

[3] Bach DS. Echo/Doppler evaluation of hemodynamics after aortic valve replacement: principles of interrogation and evaluation of high gradients. JACC Cardiovasc Imaging. 2010;3(3):296-304.

[4] Garcia D, Pibarot P, Dumesnil JG, Sakr F, Durand L-G. Assessment of Aortic Valve Stenosis Severity A New Index Based on the Energy Loss Concept. Circulation. 2000;101: 765-771.

[5] Rizk J. 4D flow MRI applications in congenital heart disease. Eur Radiol. 2021 Feb;31(2):1160-1174.

[6] Dyverfeldt P, Bissell M, Barker AJ, Bolger AF, Carlhäll C-J, Ebbers T, et al. 4D flow cardiovascular magnetic resonance consensus statement. J Cardiovasc Magn Reson. 2015;17(1):72.

[7] Stankovic Z, Allen BD, Garcia J, Jarvis KB, Markl M. 4D Flow Imaging with MRI. Cardiovasc Diagn Ther. 2014;4(2):173-192.

[8] Zhong L, Schrauben EM, Garcia J, Uribe S, Grieve SM, Elbaz MSM, et al. Intracardiac 4D Flow MRI in Congenital Heart Disease: Recommendations on Behalf of the ISMRM Flow & Motion Study Group. J Magn Reson Imaging. 2019;50(3):677-681.

[9] Pelc NJ, Herfkens RJ, Shimakawa A, Enzmann DR. Phase contrast cine magnetic resonance imaging. Magn Reson Q. 1991;7(4):229-254.

[10] Wu W, Budovec J, Foley WD. Prospective and retrospective ECG gating for thoracic CT angiography: a comparative study. AJR Am J Roentgenol. 2009;193(4):955-963.

[11] Moghari MH, Roujol S, Chan RH, Hong SN, Bello N, Henningsson M, et al. Free-breathing 3D cardiac MRI using iterative image-based respiratory motion correction. Magn Reson Med. 2013;70(4):1005-1015.

[12] Baltes C, Kozerke S, Hansen MS, Pruessmann KP, Tsao J, Boesiger P. Accelerating cine phase-contrast flow measurements using k-t BLAST and k-t SENSE. Magn Reson Med. 2005;54(6):1430-1438.

[13] Lustig M, Donoho D, Pauly JM. Sparse MRI: The application of compressed sensing for rapid MR imaging. Magn Reson Med. 2007;58(6):1182-1195.

[14] Stadlbauer A, van der Riet W, Crelier G, Salomonowitz E. Accelerated time-resolved three-dimensional MR velocity mapping of blood flow patterns in the aorta using SENSE and k-t BLAST. Eur J Radiol. 2010;75(1):e15-e21.

[15] Keller EJ, Collins JD, Rigsby C, Carr JC, Markl M, Schnell S. Superior Abdominal 4D Flow MRI Data Consistency with Adjusted Preprocessing Workflow and Noncontrast Acquisitions. Acad Radiol. 2017;24(3):350-358.

[16] Yilmaz P, Wallecan K, Kristanto W, Aben JP, Moelker A. Evaluation of a Semi-automatic Right Ventricle Segmentation Method on Short-Axis MR Images. J Digit Imaging. 2018 Oct;31(5):670-679.

[17] Avendi MR, Kheradvar A, Jafarkhani H. A combined deep-learning and deformable-model approach to fully automatic segmentation of the left ventricle in cardiac MRI. Med Image Anal. 2016;30:108-119.

[18] Cong J, Zheng Y, Xue W, Cao B, Li S. MA-Shape: Modality Adaptation Shape Regression for Left Ventricle Segmentation on Mixed MR and CT Images. IEEE Access. 2019;7: 16584-16593.

[19] Huang S, Liu J, Lee LC, Venkatesh SK, Teo LLS, Au C, et al. An image-based comprehensive approach for automatic segmentation of left ventricle from cardiac short axis cine MR images. J Digit Imaging. 2011;24(4):598-608.

[20] Litjens G, Ciompi F, Wolterink JM, de Vos BD, Leiner T, Teuwen J, et al. State-of-the-Art Deep Learning in Cardiovascular Image Analysis. JACC Cardiovasc Imaging. 2019;12(8): 1549-1565.

[21] Ronneberger O, Fischer P, Brox T. U-Net: Convolutional Networks for Biomedical Image Segmentation. In 2015. p. 234-241.

[22] Berhane H, Scott M, Elbaz M, Jarvis K, McCarthy P, Carr J, et al. Fully automated 3D aortic segmentation of 4D flow MRI for hemodynamic analysis using deep learning. Magn Reson Med. 2020;84(4):2204-2218.

[23] Bratt A, Kim J, Pollie M, Beecy AN, Tehrani NH, Codella N, et al. Machine learning derived segmentation of phase velocity encoded cardiovascular magnetic resonance for fully automated aortic flow quantification. J Cardiovasc Magn Reson. 2019;21(1):1

[24] Wu B, Fang Y, Lai X. Left ventricle automatic segmentation in cardiac MRI using a combined CNN and U-net approach. Comput Med Imaging Graph. 2020;82:101719.

[25] Medicine I of. Cardiovascular disability: Updating the social security listings. Cardiovascular Disability: Updating the Social Security Listings. National Academies Press; 2010. 1-278 p.

[26] Therrien J, Webb G. Clinical update on adults with congenital heart disease. Lancet. 2003;362(9392):1305-1313.

[27] Warmerdam E, Krings GJ, Leiner T, Grotenhuis HB. Three-dimensional and four-dimensional flow assessment in congenital heart disease. Heart. 2020;106(6):421-426.

[28] Ota H, Higuchi S, Sun W, Ueda T, Takase K, Tamura H. Four-Dimensional Flow Magnetic Resonance Imaging for Cardiovascular Imaging: from Basic Concept to Clinical Application. Cardiovasc Imaging Asia. 2018;2(2): 85-96.

[29] Jeong D, Anagnostopoulos P V., Roldan-Alzate A, Srinivasan S, Schiebler ML, Wieben O, et al. Ventricular kinetic energy may provide a novel noninvasive way to assess ventricular performance in patients with repaired tetralogy of Fallot. J Thorac Cardiovasc Surg. 2015;149(5):1339-1347.

[30] Jamalidinan F, Hassanabad AF, François CJ, Garcia J. Four-dimensional-flow Magnetic Resonance Imaging of the Aortic Valve and Thoracic Aorta. Radiol Clin North Am. 2020;58(4): 753-763.

[31] Leidenberger T, Gordon Y, Farag M, Delles M, Fava Sanches A, Fink MA, et al. Imaging-Based 4D Aortic Pressure Mapping in Marfan Syndrome Patients: A Matched Case-Control Study. Ann Thorac Surg. 2020;109(5):1434-1440.

[32] Chelu RG, van den Bosch AE, van Kranenburg M, Hsiao A, van den Hoven AT, Ouhlous M, et al. Qualitative

grading of aortic regurgitation: a pilot study comparing CMR 4D flow and echocardiography. Int J Cardiovasc Imaging. 2016 Feb;32(2):301-307.

[33] Bock J, Frydrychowicz A, Lorenz R, Hirtler D, Barker AJ, Johnson KM, et al. In vivo noninvasive 4D pressure difference mapping in the human aorta: phantom comparison and application in healthy volunteers and patients. Magn Reson Med. 2011;66(4): 1079-1088.

[34] Tyszka JM, Laidlaw DH, Asa JW, Silverman JM. Three-dimensional, time-resolved (4D) relative pressure mapping using magnetic resonance imaging. J Magn Reson Imaging. 2000;12(2):321-329.

[35] Ebbers T, Wigström L, Bolger AF, Wranne B, Karlsson M. Noninvasive measurement of time-varying three-dimensional relative pressure fields within the human heart. J Biomech Eng. 2002;124(3):288-293.

[36] Hassanabad AF, Burns F, Bristow MS, Lydell C, Howarth AG, Heydari B, et al. Pressure drop mapping using 4D flow MRI in patients with bicuspid aortic valve disease: A novel marker of valvular obstruction. Magn Reson Imaging. 2020;65:175-182.

[37] Nkomo VT, Gardin JM, Skelton TN, Gottdiener JS, Scott CG, Enriquez-Sarano M. Burden of valvular heart diseases: a population-based study. Lancet. 2006;368(9540): 1005-1011.

[38] Nishimura RA, Vahanian A, Eleid MF, Mack MJ. Mitral valve disease--current management and future challenges. Lancet. 2016;387(10025):1324-1334.

[39] Ling LH, Enriquez-Sarano M, Seward JB, Tajik AJ, Schaff H V., Bailey KR, et al. Clinical outcome of mitral regurgitation due to flail leaflet. N Engl J Med. 1996;335(19):1417-1423.

[40] Kagiyama N, Shrestha S. Echocardiographic assessment of mitral regurgitation. J Med Ultrason (2001). 2020;47(1):59-70.

[41] Enriquez-Sarano M, Akins CW, Vahanian A. Mitral regurgitation. Lancet. 2009;373(9672):1382-1394.

[42] Heitner J, Bhumireddy GP, Crowley AL, Weinsaft J, Haq SA, Klem I, et al. Clinical application of cine-MRI in the visual assessment of mitral regurgitation compared to echocardiography and cardiac catheterization. PLoS One. 2012;7(7):e40491.

[43] Chew PG, Bounford K, Plein S, Schlosshan D, Greenwood JP. Multimodality imaging for the quantitative assessment of mitral regurgitation. Quant Imaging Med Surg. 2018;8(3):342-359.

[44] Fidock B, Barker N, Balasubramanian N, Archer G, Fent G, Al-Mohammad A, et al. A Systematic Review of 4D-Flow MRI Derived Mitral Regurgitation Quantification Methods. Front Cardiovasc Med. 2019;6:103.

[45] Calkoen EE, Elbaz MSM, Westenberg JJM, Kroft LJM, Hazekamp MG, Roest AAW, et al. Altered left ventricular vortex ring formation by 4-dimensional flow magnetic resonance imaging after repair of atrioventricular septal defects. J Thorac Cardiovasc Surg. 2015;150(5): 1233-40.e1.

[46] Kräuter C, Reiter U, Reiter C, Nizhnikava V, Masana M, Schmidt A, et al. Automated mitral valve vortex ring extraction from 4D-flow MRI. Magn Reson Med. 2020;84(6): 3396-3408.

[47] Calkoen E, de Koning PJ, van der Geest RJ, de Roos A, Westenberg JJ, Roest A. Vortex flow in the left atrium in healthy controls and patients with mitral valve regurgitation after atrioventricular septal defect correction: evaluation with 4D Flow MRI and particle tracing. J Cardiovasc Magn Reson. 2015 Dec 3;17(S1):Q123.

[48] El Sabbagh A, Reddy YN V, Nishimura RA. Mitral Valve Regurgitation in the Contemporary Era: Insights Into Diagnosis, Management, and Future Directions. JACC Cardiovasc Imaging. 2018;11(4):628-643.

[49] Le Goffic C, Toledano M, Ennezat P-V, Binda C, Castel A-L, Delelis F, et al. Quantitative Evaluation of Mitral Regurgitation Secondary to Mitral Valve Prolapse by Magnetic Resonance Imaging and Echocardiography. Am J Cardiol. 2015;116(9):1405-1410.

[50] January CT, Wann LS, Alpert JS, Calkins H, Cigarroa JE, Cleveland JC, et al. 2014 AHA/ACC/HRS guideline for the management of patients with atrial fibrillation: a report of the American College of Cardiology/American Heart Association Task Force on Practice Guidelines and the Heart Rhythm Society. J Am Coll Cardiol. 2014;64(21):e1-76.

[51] Andrade JG, Aguilar M, Atzema C, Bell A, Cairns JA, Cheung CC, et al. The 2020 Canadian Cardiovascular Society/Canadian Heart Rhythm Society Comprehensive Guidelines for the Management of Atrial Fibrillation. Can J Cardiol. 2020;36(12):1847-1948.

[52] Pamukcu B, Lip GYH, Lane DA. Simplifying stroke risk stratification in atrial fibrillation patients: Implications of the CHA2DS2-VASc risk stratification scores. Age Ageing. 2010;39(5):533-535.

[53] Fluckiger JU, Goldberger JJ, Lee DC, Ng J, Lee R, Goyal A, et al. Left atrial flow velocity distribution and flow coherence using four-dimensional FLOW MRI: a pilot study investigating the impact of age and Pre- and Postintervention atrial fibrillation on atrial hemodynamics. J Magn Reson Imaging. 2013;38(3):580-587.

[54] Markl M, Lee DC, Ng J, Carr M, Carr J, Goldberger JJ. Left Atrial 4-Dimensional Flow Magnetic Resonance Imaging: Stasis and Velocity Mapping in Patients With Atrial Fibrillation. Invest Radiol. 2016;51(3):147-154.

[55] Lee DC, Markl M, Ng J, Carr M, Benefield B, Carr JC, et al. Three-dimensional left atrial blood flow characteristics in patients with atrial fibrillation assessed by 4D flow CMR. Eur Hear J – Cardiovasc Imaging. 2016;17(11):1259-1268.

[56] Markl M, Lee DC, Furiasse N, Carr M, Foucar C, Ng J, et al. Left Atrial and Left Atrial Appendage 4D Blood Flow Dynamics in Atrial Fibrillation. Circ Cardiovasc Imaging. 2016;9(9):e004984.

[57] Garcia J, Sheitt H, Bristow MS, Lydell C, Howarth AG, Heydari B, et al. Left atrial vortex size and velocity distributions by 4D flow MRI in patients with paroxysmal atrial fibrillation: Associations with age and CHA2 DS2 -VASc risk score. J Magn Reson Imaging. 2020;51(3):871-884.

[58] Kim H, Sheitt H, Jamalidinan F, Wilton S, White J, Garcia J. Left Ventricular Flow Analysis in Atrial Fibrillation. In: 2020 42nd Annual International Conference of the IEEE Engineering in Medicine & Biology Society (EMBC). IEEE; 2020. p. 1182-1185.

[59] Demirkiran A, Amier RP, Hofman MBM, van der Geest RJ, Robbers LFHJ, Hopman LHGA, et al.

Altered left atrial 4D flow characteristics in patients with paroxysmal atrial fibrillation in the absence of apparent remodeling. Sci Rep. 2021;11(1):5965.

[60] Spartera M, Pessoa-Amorim G, Stracquadanio A, Von Ende A, Fletcher A, Manley P, et al. Left atrial 4D flow cardiovascular magnetic resonance: a reproducibility study in sinus rhythm and atrial fibrillation. J Cardiovasc Magn Reson. 2021 Dec 22;23(1):29.

[61] Nistri S, Basso C, Marzari C, Mormino P, Thiene G. Frequency of bicuspid aortic valve in young male conscripts by echocardiogram. Am J Cardiol. 2005;96(5):718-721.

[62] Tutar E, Ekici F, Atalay S, Nacar N. The prevalence of bicuspid aortic valve in newborns by echocardiographic screening. Am Heart J. 2005;150(3): 513-515.

[63] Basso C, Boschello M, Perrone C, Mecenero A, Cera A, Bicego D, et al. An echocardiographic survey of primary school children for bicuspid aortic valve. Am J Cardiol. 2004;93(5):661-663.

[64] Sievers H-H, Schmidtke C. A classification system for the bicuspid aortic valve from 304 surgical specimens. J Thorac Cardiovasc Surg. 2007;133(5):1226-1233.

[65] Masri A, Svensson LG, Griffin BP, Desai MY. Contemporary natural history of bicuspid aortic valve disease: a systematic review. Heart. 2017;103(17):1323-1330.

[66] Linda AP, Thomas TT, Eric MI, others. Aortic diameter≥ 5.5 cm is not a good predictor of type a aortic dissection: observations from the International Registry of Acute Aortic Dissection (IRAD). Circulation. 2007;116(10):1120-1127.

[67] Borger MA, Fedak PWM, Stephens EH, Gleason TG, Girdauskas E, Ikonomidis JS, et al. The American Association for Thoracic Surgery consensus guidelines on bicuspid aortic valve–related aortopathy: Full online-only version. J Thorac Cardiovasc Surg. 2018;156(2):e41–e74.

[68] Rodríguez-Palomares JF, Dux-Santoy L, Guala A, Kale R, Maldonado G, Teixidó-Turà G, et al. Aortic flow patterns and wall shear stress maps by 4D-flow cardiovascular magnetic resonance in the assessment of aortic dilatation in bicuspid aortic valve disease. J Cardiovasc Magn Reson. 2018;20(1):28.

[69] Mahadevia R, Barker AJ, Schnell S, Entezari P, Kansal P, Fedak PWM, et al. Bicuspid aortic cusp fusion morphology alters aortic three-dimensional outflow patterns, wall shear stress, and expression of aortopathy. Circulation. 2014;129(6):673-682.

[70] Garcia J, Barker AJ, Collins JD, Carr JC, Markl M. Volumetric quantification of absolute local normalized helicity in patients with bicuspid aortic valve and aortic dilatation. Magn Reson Med. 2017;78(2):689-701.

[71] Dux-Santoy L, Guala A, Teixidó-Turà G, Ruiz-Muñoz A, Maldonado G, Villalva N, et al. Increased rotational flow in the proximal aortic arch is associated with its dilation in bicuspid aortic valve disease. Eur Hear Journal-Cardiovascular Imaging. 2019;20(12):1407-1417.

[72] Fatehi Hassanabad A, Garcia J, Verma S, White JA, Fedak PWM. Utilizing wall shear stress as a clinical biomarker for bicuspid valve-associated aortopathy. Curr Opin Cardiol. 2019;34(2):124-131.

[73] Bissell MM, Hess AT, Biasiolli L, Glaze SJ, Loudon M, Pitcher A, et al. Aortic dilation in bicuspid aortic valve disease: flow pattern is a major contributor and differs with valve fusion type. Circ Cardiovasc Imaging. 2013;6(4):499-507.

[74] Meierhofer C, Schneider EP, Lyko C, Hutter A, Martinoff S, Markl M, et al. Wall shear stress and flow patterns in the ascending aorta in patients with bicuspid aortic valves differ significantly from tricuspid aortic valves: a prospective study. Eur Heart J Cardiovasc Imaging. 2013;14(8):797-804.

[75] van Ooij P, Markl M, Collins JD, Carr JC, Rigsby C, Bonow RO, et al. Aortic Valve Stenosis Alters Expression of Regional Aortic Wall Shear Stress: New Insights From a 4-Dimensional Flow Magnetic Resonance Imaging Study of 571 Subjects. J Am Heart Assoc. 2017;6(9).

[76] Guzzardi DG, Barker AJ, Van Ooij P, Malaisrie SC, Puthumana JJ, Belke DD, et al. Valve-related hemodynamics mediate human bicuspid aortopathy: insights from wall shear stress mapping. J Am Coll Cardiol. 2015;66(8):892-900.

[77] Lindroos M, Kupari M, Heikkilä J, Tilvis R. Prevalence of aortic valve abnormalities in the elderly: An echocardiographic study of a random population sample. J Am Coll Cardiol. 1993;21(5):1220-1225.

[78] Vahanian A, Baumgartner H, Bax J, Butchart E, Dion R, Filippatos G, et al. Guidelines on the management of valvular heart disease: The Task Force on the Management of Valvular Heart Disease of the European Society of Cardiology. Eur Heart J. 2006;28(2): 230-268.

[79] Nishimura RA, Otto CM, Bonow RO, Carabello BA, Erwin JP, Fleisher LA, et al. 2017 AHA/ACC Focused Update of the 2014 AHA/ACC Guideline for the Management of Patients With Valvular Heart Disease: A Report of the American College of Cardiology/American Heart Association Task Force on Clinical Practice Guidelines. J Am Coll Cardiol. 2017;70(2):252-289.

[80] Minners J, Allgeier M, Gohlke-Baerwolf C, Kienzle R-P, Neumann F-J, Jander N. Inconsistent grading of aortic valve stenosis by current guidelines: haemodynamic studies in patients with apparently normal left ventricular function. Heart. 2010;96(18):1463-1468.

[81] Dyverfeldt P, Hope MD, Tseng EE, Saloner D. Magnetic resonance measurement of turbulent kinetic energy for the estimation of irreversible pressure loss in aortic stenosis. JACC Cardiovasc Imaging. 2013;6(1):64-71.

[82] Barker AJ, van Ooij P, Bandi K, Garcia J, Albaghdadi M, McCarthy P, et al. Viscous energy loss in the presence of abnormal aortic flow. Magn Reson Med. 2014 Sep;72(3):620-628.

[83] Binter C, Gotschy A, Sündermann SH, Frank M, Tanner FC, Lüscher TF, et al. Turbulent Kinetic Energy Assessed by Multipoint 4-Dimensional Flow Magnetic Resonance Imaging Provides Additional Information Relative to Echocardiography for the Determination of Aortic Stenosis Severity. Circ Cardiovasc Imaging. 2017;10(6).

[84] Garcia J, Barker AJ, Murphy I, Jarvis K, Schnell S, Collins JD, et al. Four-dimensional flow magnetic resonance imaging-based characterization of aortic morphometry and haemodynamics: impact of age, aortic diameter, and valve morphology. Eur Hear J – Cardiovasc Imaging. 2016;17(8):877-884.

[85] Hope MD, Sigovan M, Wrenn SJ, Saloner D, Dyverfeldt P. MRI hemodynamic markers of progressive bicuspid aortic valve-related aortic disease. J Magn Reson Imaging. 2014;40(1):140-145.

[86] Baumgartner H, Bonhoeffer P, De Groot NMS, de Haan F, Deanfield JE, Galie N, et al. ESC Guidelines for the management of grown-up congenital heart disease (new version 2010). Eur Heart J. 2010;31(23):2915-2957.

[87] Teo LLS, Cannell T, Babu-Narayan S V., Hughes M, Mohiaddin RH. Prevalence of associated cardiovascular abnormalities in 500 patients with aortic coarctation referred for cardiovascular magnetic resonance imaging to a tertiary center. Pediatr Cardiol. 2011;32(8):1120-1127.

[88] Hope MD, Meadows AK, Hope T a, Ordovas KG, Saloner D, Reddy GP, et al. Clinical evaluation of aortic coarctation with 4D flow MR imaging. J Magn Reson Imaging. 2010;31(3):711-718.

[89] Riesenkampff E, Fernandes JF, Meier S, Goubergrits L, Kropf S, Schubert S, et al. Pressure fields by flow-sensitive, 4D, velocity-encoded CMR in patients with aortic coarctation. JACC Cardiovasc Imaging. 2014;7(9):920-926.

[90] Frydrychowicz A, Markl M, Hirtler D, Harloff A, Schlensak C, Geiger J, et al. Aortic hemodynamics in patients with and without repair of aortic coarctation: in vivo analysis by 4D flow-sensitive magnetic resonance imaging. Invest Radiol. 2011;46(5):317-325.

Chapter 4

Cardiovascular Changes during Robot-Assisted Pelvic Surgery

Ildar I. Lutfarakhmanov, Peter I. Mironov, Ildar R. Galeev and Valentin N. Pavlov

Abstract

The application of robotic assistance in pelvic surgery has become popular across multiple specialties during the past decades, facilitating minimally invasive surgery. The most remarkable challenges regarding these procedures are the carbon dioxide pneumoperitoneum and steep Trendelenburg position. The combination of two factors affects the patient additionally or synergistically and have important physiological effects on cardiovascular system. All those changes are usually well tolerated in patients with normal cardiac function, but it can be different in elderly patients or even in patients with underlying heart conditions. In order to provide the proper management of patients undergone the robotic surgery, we aim to thoroughly understand these effects and overview the risks and possible related cardiovascular complications. Further, a short introduction on dangerous areas of robot-assisted pelvic surgery will be briefly reviewed.

Keywords: robotic surgery, pneumoperitoneum, Trendelenburg position, cardiovascular system, central hemodynamics

1. Introduction

Prostate cancer remains the most common solid organ malignancy and the second leading cause of cancer death in the US men [1]. Radical prostatectomy is a standard treatment option for localized carcinoma of the prostate, that showed a significant relative risk reduction in cancer-specific mortality as compared with watchful waiting. Radical cystectomy and pelvic lymph node dissection are the standard treatment options for muscle-invasive bladder carcinoma, but this operation is a surgical procedure associated with the highest morbidity and mortality among all surgical operations. Laparoscopic cystectomy was introduced to decrease associated morbidity, and resulted in a significantly lower intra-operative blood loss and transfusion rates, lower pain scores, and allowing a more rapid resumption of oral intake and a shorter hospital stay. Uterine cancer is one of the few cancers with increasing incidence and mortality in the United States. It is the fourth most common cancer diagnosed and the seventh most common cause of cancer death among US women [2]. Hysterectomy is the second most commonly performed procedure in women of reproductive age, and nearly 600,000 hysterectomies are performed annually; about 20 million US women have had a hysterectomy [3]. In 2016, colorectal cancers accounted for 8.5% of all new malignant cancer cases and 8.7%

of all cancer deaths in the US [4]. Laparoscopic approach has become the preferred standard of care in colorectal surgery and has been proven to be as safe and effective as open surgery, and associated with a lower blood loss and shorter length of stay.

The introduction of robot-assisted surgery, specifically the da Vinci Surgical System (Intuitive Surgical, Inc), is one of the biggest breakthroughs in past decades, and represents the most significant advancement in minimally invasive surgery. Initially used in urology, particularly in radical prostatectomy, the robotic technique became the most suitable for operations in a closed and restricted pelvic space across different specialties including colorectal and gynecologic surgery, making the indications for robot-assisted laparoscopic pelvic surgery (RALPS) widely increased. Robotic assistance allows a three-dimensional stable operator-controlled high-definition magnified view of the operative field, filtrating the tremor of the surgeon, and enabling precise movements of instruments with seven degrees of freedom, overtaken the limitations of standard open and/or laparoscopic surgery. Robot-assisted surgery represents an advantage in terms of smaller incisions, precise dissection in a confined space, reduction of intraoperative blood loss and lower transfusion rate, fewer postoperative pain and complications, decreased in-hospital death rate and length of stay, and faster return to daily functions. Age and obesity are significant risk factors for malignancy, yet complicate the surgery. RALPS is feasible and safe for patients with complicated diseases compared with conventional laparoscopic or open surgeries. Robotic procedures can reduce the cost of operation in elderly or morbidly obese patients with comorbidities, as these procedures generally require a shorter hospital stay than open methods. As the prevalence of obesity climbs steadily, the treatment of these patients with robotic surgery will increase. Multiple postoperative advantages of this technique permit safe management of patients with more and more severe cardiorespiratory disease.

As the previous surgical techniques, this one is not without limitations. First, potentially long duration of surgery (initial case series reported a longer mean operative time up to 270 minutes but this has since been reduced to a mean of 150 minutes). Second, RALPS requires a much Trendelenburg tilt of 30 to 45 degrees – steep Trendelenburg position (sTrP). This extreme position causes significant and potentially adverse cardiovascular alterations. The most remarkable challenge regarding RALPS is the carbon dioxide (CO_2) pneumoperitoneum (PP) resulting in ventilatory and respiratory changes. In compromised patients, cardiorespiratory disturbances aggravate the hypercapnia. Gasless laparoscopy may be helpful to reduce the pathophysiological changes induced by PP but increases technical difficulty. This becomes more complicated in operations where sTrP and PP are combined, increasing the risk of hemodynamic disorders.

2. Pathophysiological cardiovascular effects of steep Trendelenburg position and pneumoperitoneum during robot-assisted pelvic surgery

In RALPS, hemodynamic alterations are derived from the three origins: first is the intraabdominal pressure (IAP) created by the PP; second is the existence of an insufflation CO_2 that is absorbed by the blood; third is the sTrP of the patient. These factors account for the majority of physiological changes in the patient in addition to the events that develop from surgical intervention. The interactions of these factors are important in the intraoperative management of the RALPS patient. Standard IAP levels between 12 and 15 mm Hg cause increase in cardiac output by applying pressure to splanchnic venous bed and are useful for performing procedures in most cases [5]. On the other hand, higher insufflation pressure up to 20 mm Hg apply compression over inferior cava vena causing a preload reduction and a decrease in the cardiac output, but are safe and have no significant short-term effects on organ perfusion, and not associated with a higher complication rate [6]. Steep TrP is essential for the final hemodynamic result during RALPS; it increases the venous return and compensates blood loss. The combination of two factors the sTrP and PP can cause significant cardiovascular changes, and an increase in the angle of inclination can further exacerbate these changes [7]. PP with high IAP and sTrP may affect the excretion of CO_2. If pulmonary ventilation is not enough to eliminate the CO_2 absorbed from the PP, the dissolved CO_2 causes hypercapnia and acidosis. Hypercapnia and acidosis decrease the cardiac contractility, make myocardium more sensitive to catecholamines and cause peripheral vasodilatation. Sympathetic activation caused by hypercapnia finally leads to tachyarrhythmia and peripheral vasoconstriction [8]. Vagal (parasympathetic) stimulation may cause bradyarrhythmia in a range from bradycardia to asystole, and hypotension [9]. These effects may further complicate clinical management of patients with underlying chronic lung disease or the morbidly obese. Technical disadvantages of RALPS in patients with obesity include excessive fat tissue, deeper and narrowed true pelvis that result in limited working space, long distance to operative field, difficulty in trocars placement, and suboptimal visualization [10, 11].

Table 1 shows the results of different studies in which the intraoperative hemodynamics, cardiac function, filling volume and pressure, and cardiac variations were measured through the operation time relative to the values before the induction of anesthesia. Studies reporting mean arterial pressure (MAP) changes have not been consistent. In a majority of studies, an increased or unaffected MAP was demonstrated. In the most of studies there were no changes in heart rate (HR), whereas one study showed an increase, and the remaining studies showed decreased HR. In general, the stroke volume (SV), cardiac output (CO), and cardiac index (CI) remained stable, except one study, in which cardiac function significantly decreased, authors explained by the combined effect of general anesthesia and PP. All the reported studies described an increased or unchanged systemic vascular resistance (SVR). SVR also increased in studies in which no decrease in CO was reported. Whereas the normal heart tolerates increases in afterload under physiologic conditions, the changes in afterload produced by the PP can result in deleterious effects in patients with cardiac diseases and may lead to further decrease in CO. Significant 80% [34] up to 200% [43] increasing in Central venous pressure (CVP) as well as pulmonary mean artery pressure (PMAP) and pulmonary artery wedge pressure (PAWP) after the institution of PP and sTrP was seen in all of studies.

2.1 Safe effects of steep Trendelenburg position and pneumoperitoneum on central hemodynamics

Pneumoperitoneum resulted in an increased MAP, decreased SV, and increased pulse pressure variation (PPV) and stroke volume variation (SVV) compared with the supine position. Pneumoperitoneum combined with sTrP resulted in an unchanged MAP, increased SV, unchanged PPV, and increased SVV compared with isolated PP in patients without cardiopulmonary disease. The PP induced increases in abdominal pressure may compress the inferior vena cava, resulting in decreased right ventricular preload [14].

The effects of extraperitoneal and transperitoneal CO_2 insufflation on hemodynamic parameters were assessed in [17]. HR did not change significantly during either of the surgical approaches. Although MAP and CO did increase significantly during transperitoneal insufflation, although not during extraperitoneal insufflation, the differences between the methods were not statistically significant.

Function	Status		
	Increased	**Decreased**	**Unchanged**
Hemodynamics			
Mean arterial pressure	[7, 12–25]	[26–33]	[34–42]
Heart rate	[25]	[14, 23, 26, 27, 29–31, 33, 36]	[7, 12, 13, 15–22, 24, 28, 34, 35, 38–40, 42]
Cardiac function			
Stroke volume	[32]	[27]	[19, 22, 24, 38]
Cardiac output	—	[26, 27]	[17, 19, 20, 22, 24, 38, 41]
Cardiac index	—	[27]	[15, 16, 18, 24, 25, 29, 34]
Ventricular end-diastolic area / stroke work index / ejection fraction	[20, 22]	[24]	—
Systemic vascular resistance	[19]	—	[22, 25, 29, 34, 38]
Filling volume / pressure			
Aortic diameter	[19]	—	—
Central venous pressure	[12, 15, 16, 18, 20–23, 25, 27, 34, 35, 37, 38, 40]	—	—
Pulmonary mean artery pressure	[15, 22]	—	—
Pulmonary artery wedge pressure	[15, 22]	—	—
Cardiac variations			
Heart rate[a]	[36]	—	—
QTc interval[b]	[28]	—	—
Tp-e interval[c]	—	—	[28]
Pulse pressure	[14]	—	—
Stroke volume	[14]		[20]

[a]*Computerized ECG.*
[b]*HR-corrected QT (QTc) interval.*
[c]*Tpeak–Tend (Tp-e) interval.*

Table 1.
Summary of literature search, including changes in intraoperative hemodynamics, cardiac function, filling volume and pressure, and cardiac variations during RALPS.

Despite a 50–100% increasing in CVP, PMAP and PAWP, the sTrP and PP did not change the CI and contractility of the right ventricle [15].

Cerebral blood flow-carbon dioxide reactivities in the supine and modest TrP under PP were compared in [16]. The main result is that there is no change between the supine position and the 30° TrP.

Combination of the 45° sTrP and insufflation pressures of 20 mm Hg resulted in a MAP reduced by 17%, HR reduced by 21%, and CO reduced by 37%. Overall, patients tolerated the procedure well with minimal clinically significant cardio-pulmonary effects. For patients with limited cardiopulmonary reserve, however, physicians must weight these benefits with the negative cardiovascular changes associated with this type of procedure [26].

Steep TrP and high-pressure PP leads to significant decrease in SV and CO. Although hemodynamic parameters decreased compared to the baseline, they were within the physiological normal limits and all these parameters returned to baseline after deflation of PP in the supine position [27].

The effects of volatile anesthetic sevoflurane and intravenous propofol on Central hemodynamics were compared in [18]. No intergroup differences in MAP, CI, and CVP being at any time point in PP and sTrP, even though HR was lower in sevoflurane group. Compared with time point before surgery, MAP and CVP were significantly higher in both groups. There were no differences within each group for CI at each time point. There were no major complications on normal postoperative rounds in either group.

Steep TrP significantly increased the SV, whereas PP decreased the aortic diameter, and the combination of sTrP and PP significantly increased MAP and CVP, but did not change CO and SV [19].

Improved hemodynamics with significant increase of MAP, CVP and CO under combination sTrP with PP were observed in [20]. All the variables studied remained within the clinically acceptable range and did not deteriorate left or right ventricular function.

Central venous pressure increased after the institution of PP and sTrP and returned to baseline range following reinstitution of supine position after completion of robotic pelvic surgeries. These findings can be explained by combination of increased intra-abdominal, intrathoracic pressures and acute volume loading during PP and sTrP [35].

Although the sTrP combined with a PP significantly influenced cardiovascular homeostasis, all investigated variables remained within a clinically acceptable range, and a combination of the prolonged sTrP and PP was well tolerated by patients. During institution of the sTrP, MAP and CVP increased significantly. The observed increase in pressure is the result of increased hydrostatic pressure at the external auditory meatus caused by the tilting of the table. Because MAP increased by a greater absolute amount than CVP (34 vs. 23 mm Hg, respectively), at least part of the increase in MAP must also be caused by increased CO, SVR, or both [21].

The influence of sTrP on cerebral hemodynamic homeostasis was elucidated in [37]. While patients were in the sTrP, zero flow pressure (ZFP) increased in parallel with CVP. The pulsatility and resistance indexes, and ZFP – CVP gradient did not increase significantly after 3 h of the sTrP. Albeit the unphysiological sTrP combined with the need for PP raises major concerns for the physiological homeostasis of the patient, especially for intracranial pressure and brain perfusion, it does not compromise cerebral perfusion and seems to be well tolerated by most patients.

RALPS performed with low pressure (8 mm Hg) PP and sTrP was associated with significant variations in arterial pressure, CVP, SVR, left ventricular end-diastolic and end-systolic volume, and ejection fraction. With the return to neutral position at the end of surgery, with PP deflation, most of the assessed parameters returned to baseline value, with the exception of SVR and CO. However, all variables remained within limits safely manageable by the anaesthesiologists [38].

Two- to three-fold increases of right- as well as left-sided filling pressures during PP with sTrP was found in [22]. The index of left ventricular filling was at a level seen during heart failure. Pulmonary hypertension, with systolic pulmonary artery pressure exceeding 35 mm Hg, was recorded in 75% of the patients. Furthermore, right-sided and left-sided filling pressures were almost equal. Systemic blood pressure was also increased, but there was no change in CO.

Hemodynamic changes, such as a significant decrease in HR after induction of anesthesia, the insufflation of PP and the transfer patient to sTrP were found in [29].

In the time of operation, the significant decrease of MAP was also observed. Although these hemodynamic parameters were reduced compared to the baseline level, they were within the physiological norm and all indicators returned to the baseline level after the elimination of PP. There were no cases of cardiovascular complications in the early postoperative period.

Low-pressure PP, sTrP and mechanical ventilation may interact and be the origin of the alterations in the autonomic nervous system modulation of HR variability. Minor alterations in cerebral oxygenation and autonomic perturbations do not cause clinically significant alterations in a patient's HR variability. This finding supports the safety of RALPS [36].

The impact of overweight on hemodynamic in patients undergoing RALPS with PP was investigated in [40]. The creation of prolonged PP had no adverse effects on hemodynamic parameters and no clinically relevant cardiovascular change was noted.

Cardiac output was not affected either by the sTrP nor establishment of a PP. Head-down position and prolonged PP were associated with an early elevation in CVP that was sustained and did not vary. A temporary increase of MAP was recorded. After the release of the PP, hemodynamic parameters returned to baseline levels except HR and CI [34].

Changes in circulatory status by measuring hemodynamic and cardiac function brought about by 28° TrP and establishment of the PP were examined in [24]. They found that head-down tilt and PP significantly decreased left ventricular ejection fraction but that left ventricular end-diastolic volume and CI did not change. These findings indicate that a sTrP and PP did not greatly influence cardiac function.

The clinical effects of general anesthesia with inhalational sevoflurane and total intravenous anesthesia with propofol were compared. Mean arterial pressure decreased in period of anesthesia induction and patient preparation and increased with sTrP and PP. Heart rate significantly decreased during the operation in both groups after sTrP and PP were established. All values were in physiologic ranges in all times [31].

Even with a standard 11–14 mm Hg PP, some form of mesenteric-splanchnic injury might be induced due to an extra gravitational pressure or a traction force caused by the exclusive fixed 45° TrP in addition to the duration of RALPS about 3.5 hours, but not any significant mesenteric-splanchnic ischemia was demonstrated [41].

Unchanged preload conditions, a slightly reduced contractility, and an 8% increase in HR, together with a 32% increase in SVR, combine to cause an 8% decrease in CO during pronounced PP, but generally normalized afterload and myocardial oxygen demand [25].

Brain regional oxygen saturation increased after the PP and further increased temporarily after the sTrP, and decreased afterwards. These changes were along with the alteration of MAP, but did not correlate with the changes of HR, indicating that MAP is the critical factor in the cerebral oxygenation. Anyway, PP and the sTrP in RALRP did not aggravate cerebral oxygenation [30].

Lower limb perfusion significantly increased after induction of anesthesia, establishment of PP, placing in TrP, when compared to the baseline in patients without previous episodes of peripheral vascular disease and morbid obesity. Arterial pressure decreased after induction of anesthesia and continued to show a decreasing trend throughout the operation. Cardiac index increased after TrP. These changes in cardiovascular physiology had negative effects on systemic perfusion, but in general, sTrP and general anesthesia improve microcirculation [32].

2.2 Negative effects of steep Trendelenburg position and pneumoperitoneum on central hemodynamics

Steep TrP is a challenging clinical setting to anesthetists due to the risks of position and long duration of PP. The creation of PP and sTrP increased MAP, that can be explained by the increase of hydrostatic pressure caused by the tilting of the table, also is caused by increased CO and SVR. PP and head-down position caused acute volume loading which causes acute elevation of CVP. Over subsequent hours, the MAP and CVP remained stable, then decreased significantly after reassuming the supine position, but remained within acceptable ranges [12].

Pneumoperitoneum and TrP caused an increase in MAP and middle ear pressure. Although the magnitude of this increase was within the normal range and none of patients suffered from ear problems postoperatively, this propensity for increase may cause problems in patients with preexisting disease [13].

The degree of head-down angle affects the cardiovascular parameters. MAP was increased significantly with PP, and showed a much greater increase (up to 31% compared with baseline) in the first 5 min after placing in the sTrP [7].

The effect of dexmedetomidine on QTc was evaluated in [28]. None of the patients had a QTc interval of >450 msec before surgery, but sTrP and PP resulted in a significant lengthening of the QTc interval > 450 msec in 2 patients and > 20 msec prolongation from baseline in 22 patients (96%). The overall result was that sTrP and PP can increase the risk of torsade de pointes for patients susceptible to ventricular arrhythmias, even when preoperative ECG findings are normal [28].

High prolonged PP and sTrP can produce adverse cardiovascular effects, and the results of study [23] demonstrated that MAP remained unchanged, but HR decreased significantly and required intervention. The CVP values were also above the normal limits. These high values might be due to the sTrP as they returned to their initial values by the end of the operation. Although the most obvious effects on HR, MAP, and CVP occurred immediately after the patients were moved into the sTrP with PP, these measurements continued to be affected to a lesser degree until the supine positioning at the end of the procedures. The most obvious changes were observed in the CVP.

Autonomic nervous system regulations following sTrP were investigated via evaluating HR variability [44]. A statistically significant decrease in the values of low-frequency and high-frequency spectral bands, representing sympathetic and parasympathetic activity, respectively, correlated with 20% decrease in HR, and in other hemodynamic parameters from the start until the end of the operation.

Significant increase in concentration of malondialdehyde regarded to be the most reliable and producible markers of oxidative stress in the clinical setting, and decrease in gastric intramucosal pH value after the induction of PP was observed in [39]. After PP deflation, these values increased steadily and reached a peak 30 minutes after deflation, which was significantly higher than that during PP insufflation. Consistent with these findings, a prolonged PP could lead to decreased splanchnic blood flow and increased oxidative stress, not only during the PP but also after the deflation.

Hemodynamic changes during RALPS reveal autonomic response to the challenges (i.e., general anesthesia and head down position), and non-neurally mediated increase of systolic arterial pressure caused by PP. Association between the vagal stimulation due to sTrP and sympathetic withdrawal caused by general anesthesia could lead to severe bradycardia and cardiac arrest in risky patients [45].

Cerebrovascular autoregulation slowly impaired over a long time period in sTrP combined with PP. High MAP might trigger or aggravate the formation of cerebral edema. To avoid neurological deterioration in patients placed in an sTrP for more than 3 h, it may be beneficial to maintain the MAP within the normal range and to minimize the duration of sTrP as much as possible [42].

The concomitant pathology, cardiac depressants, age, the duration of surgical intervention and anesthesia are of great importance to observed unfavorable changes in a cardiovascular system. The higher intra-abdominal pressure and the clearly expressed TrP create a prerequisite for more frequent hemodynamic changes, concealing a risk for the life of sick persons [33].

3. Cardiovascular complications occurring the robot-assisted laparoscopic pelvic surgery

Steep TrP and PP are required to allow adequate surgery exposure in robotic pelvic procedures. The risk of perioperative cardiovascular complications is increased by a long-time in the proper patient positioning. In contrast with respiratory complications, hemodynamic complications do not increase with surgery duration. Positioning-related complications are even more common in obese patients related to weight pressure and longer operative time. Peritoneal insufflation can result in hypotension, arrhythmias (bradycardia) or even cardiac arrest (asystole) due to vagal response, especially in patients with cardiovascular disease. We found very few cardiovascular complications in four case reports [46–49], two review articles [50, 51], five prospective [12, 20, 23, 52, 53], and ten retrospective [54–64] analyses, shown in **Table 2**.

Myocardial infarction was a result of intraoperative drug-eluting stent thrombosis after a patient developed a new left bundle branch block and was ultimately taken to the cardiac catheterization lab [47]. A patient with significant cardiac risk factors underwent a RALPS with cardiac arrest and was subsequently successfully resuscitated [46]. Another fatal myocardial infarction was in a 52-year-old patient with ASA physical status IV [48]. In a case of 72-year-old man, after 22° PP creation, severe bradycardia and complete atrioventricular block were observed, which is considered to be attributed to a vagal reflex; thus, the surgery was extended by inserting a temporary pacemaker [49].

Cardiac events	Procedure characteristic	Patient's demographics: number (n); ASA functional class; age; BMI	Number of patients (rate) with events
Arrhythmias (atrial fibrillation / bradycardia)	[12, 23, 50, 61, 63]	n = 874; ASA I-IV; 26 to 84 years; 16 to 44 kg/m^2	20 (2.3%)
Myocardial ischemia / infarction	[54, 56, 57, 59, 63]	n = 2717; ASA I-IV; 44 to 78 years; 19 to 38 kg/m^2	11 (0.4%)
Cardiac arrest	[54]	n = 1241; ASA I-IV; 46 to 74 years; overweight and obese 69.6%	4 (0.3%)
Other cardiac complications (mitral insufficiency, pulmonary edema)	[20, 52]	n = 14; ASA II-III; 24 to 61 years; 19 to 38 kg/m^2	2 (14%)
Cardiac complications not detailed (including myocardial infarction, arrhythmia, heart failure, shock)	[51, 53, 55, 58, 60, 62, 64]	n = 57220; 32 to 90 years; 17 to 70 kg/m^2	465 (0.8%)

Table 2.
Types and rates of cardiovascular complications.

Analyses of a large database demonstrated 0.9% cardiac complications after RALPS. Higher prevalence of cardiovascular-related comorbidities in morbidly obese patients may be involved in the increased incidence of cardiac complications [64]. In a largest sample study evaluated and compared incidence of perioperative complications among of non-obese, obese, and morbidly obese patients undergoing RALRP, the rates of intraoperative complications were similar [60].

4. Perioperative monitoring during robot-assisted laparoscopic pelvic surgery

RALPS is performed while the patient is under general anesthesia with endotracheal intubation. For most procedures, the standard American Society of Anesthesiologists monitoring is sufficient. These includes noninvasive blood pressure, electrocardiogram, pulse oximetry, capnography, temperature monitoring, bispectral index, and urine output. Because of relatively short operative times and minimal blood loss, invasive monitoring is rarely indicated. One freely flowing peripheral IV and plethysmography offer necessary access and hemodynamic information. In [65] non-invasive continuous arterial blood pressure measurements using the ClearSight system (BMEYE, Amsterdam, The Netherlands) were not comparable to those obtained invasively in patients undergoing RALPS because of the device tended to overestimate blood pressure. For hemodynamic optimization, stroke volume estimation and its response to fluid infusion is recommend. There is no justification for CVP nor pulmonary artery catheter monitoring based on hemodynamic changes related to PP alone.

Additional monitoring should be considered to account for patient co-morbidities, the risk of intraoperative bleeding, or longer operative times. Hemodynamically unstable or patients with cardiovascular disease intra-arterial blood pressure may be monitored by arterial cannulation [25]. The most popular additional cardiovascular monitoring for older patients with cardiopulmonary comorbidities includes:

- Invasive arterial blood pressure with cardiac output monitoring: FloTrac/Vigileo™ (Edwards Lifesciences LLC, Irvine, CA, USA) [16, 18, 27, 32]

- Transesophageal echocardiography: Haemosonic-100 (Arrow International; Everett, MA, USA); Vivid i™ (General Electrics); iE33® X7-2t (Philips N.V., Amsterdam, Netherlands); TE-V5M (Acuson Sequoia C512 Ultrasound system; Siemens, Malvern, PA) alone [19, 26] or with FloTrac/Vigileo™ [14, 20, 24, 38]

- Transpulmonary arterial thermodilution: Swan-Ganz CCO combo (CCO/SvO_2 Edwards Lifesciences LLC, Irvine, CA, USA); PiCCO Plus (Pulsion Medical Systems, Munich, Germany); Draeger Infinity Delta PiCCO SmartPod (Draeger Medical Systems, Inc., Telford, PA, USA) [15, 22, 25, 34]

As sum, Joint Consensus on Anesthesia in Urologic and Gynecologic Robotic Surgery (JC-STARS group) recommend tailored hemodynamic monitoring of the patient based on the perioperative risk [66].

5. Conclusions

Robot-assisted pelvic surgery with the da Vinci surgical system is increasingly being applied. Despite the increasing popularity, there is no unequivocal evidence to show the superiority of robotic surgery over traditional laparoscopic surgery in terms of cardiovascular complications. Interpreting the effects of the steep Trendelenburg position and that of CO_2 pneumoperitoneum separately is impossible; the combination of the factors affects the patient additionally or synergistically and have important physiological effects on cardiovascular system. All those changes are usually well tolerated in patients with normal cardiac function, but it can be different in elderly patients with ASA II-III risk or even in patients with underlying heart conditions. Cardiovascular complications not appear to be unique to RALPS and had no greater incidence. The intraoperative management of the RALPS patient presents manageable challenges. Patients should be properly monitored to understand the current situation, to maintain stability and to avoid the complications with the necessary interventions on time.

Author details

Ildar I. Lutfarakhmanov[1*], Peter I. Mironov[1], Ildar R. Galeev[1] and Valentin N. Pavlov[2]

1 Department of Anesthesiology and Intensive Care, Federal State Budgetary Educational Institution of Higher Education (Bashkir State Medical University) of Ministry of Healthcare of the Russian Federation, Ufa, Russia

2 Department of Urology, Rector, Russian Academy of Sciences, Federal State Budgetary Educational Institution of Higher Education (Bashkir State Medical University) of Ministry of Healthcare of the Russian Federation, Ufa, Russia

*Address all correspondence to: lutfarakhmanov@yandex.ru

References

[1] Siegel RL, Miller KD, Jemal A. Cancer statistics, 2019. CA: a Cancer Journal for Clinicians 2019;69(1):7-34. DOI: 10.3322/caac.21551

[2] Cronin KA, Lake AJ, Scott S, Sherman RL, Noone AM, Howlader N, Henley SJ, Anderson RN, Firth AU, Ma J, Kohler BA, Jemal A. Annual Report to the Nation on the Status of Cancer, part I: National cancer statistics. Cancer 2018;124(13):2785-2800. DOI: 10.1002/cncr.31551

[3] Hysterectomy surveillance – United States, 1994-1999 [Internet]. Available from: https://www.cdc.gov/mmwr/preview/mmwrhtml/ss5105a1.htm [Accessed: 2021-06-27]

[4] Colorectal Cancer, United States – 2007-2016 [Internet]. Available from: https://www.cdc.gov/cancer/uscs/about/data-briefs/no16-colorectal-cancer-2007-2016.htm [Accessed: 2021-06-27]

[5] Christensen CR, Maatman TK, Maatman TJ, Tran TT. Examining clinical outcomes utilizing low-pressure pneumoperitoneum during robotic-assisted radical prostatectomy. Journal of Robotic Surgery 2016;10(3):215-219. DOI: 10.1007/s11701-016-0570-3

[6] Modi PK, Kwon YS, Patel N, Dinizo M, Farber N, Zhao PT, Salmasi A, Parihar J, Ginsberg S, Ha YS, Kim IY. Safety of robot-assisted radical prostatectomy with pneumoperitoneum of 20 mmHg: a study of 751 patients. Journal of Endourology 2015;29(10):1148-1151. DOI: 10.1089/end.2015.0094

[7] Kadono Y, Yaegashi H, Machioka K, Ueno S, Miwa S, Maeda Y, Miyagi T, Mizokami A, Fujii Y, Tsubokawa T, Yamamoto K, Namiki M. Cardiovascular and respiratory effects of the degree of head-down angle during robot-assisted laparoscopic radical prostatectomy. The International Journal of Medical Robotics and Computer Assisted Surgery 2013;9(1):17-22. DOI: 10.1002/rcs.1482

[8] Chen Y, Xie Y, Xue Y, Wang B, Jin X. Effects of ultrasound-guided stellate ganglion block on autonomic nervous function during CO2-pneumoperitoneum: a randomized double-blind control trial. Journal of Clinical Anesthesia 2016;32:255-261. DOI: 10.1016/j.jclinane.2016.03.019

[9] Yong J, Hibbert P, Runciman WB, Coventry BJ. Bradycardia as an early warning sign for cardiac arrest during routine laparoscopic surgery. International Journal for Quality in Health Care 2015;27(6):473-478. DOI: 10.1093/intqhc/mzv077

[10] Bandini M, Gandaglia G, Briganti A. Obesity and prostate cancer. Current Opinion in Urology 2017;27(5):415-421. DOI: 10.1097/MOU.0000000000000424

[11] Yates J, Munver R, Sawczuk I. Robot-assisted laparoscopic radical prostatectomy in the morbidly obese patient. Prostate Cancer 2011;2011:618623. DOI: 10.1155/2011/618623

[12] Abbas DN, Kamal JM, El Sheikh SM, Mahmod AM. Early experience in anesthesia of robot assisted cystoprostatectomy. Egyptian Journal of Anaesthesia 2013;29(1):77-81. DOI: 10.1016/j.egja.2012.09.003

[13] Bozkirli F, Bedirli N, Akçabay M. Effects of steep Trendelenburg position and pneumoperitoneum on middle ear pressure in patients undergoing robotic radical prostatectomy. Turkish Journal of Medical Sciences 2017;47:295-299. DOI: 10.3906/sag-1601-113

[14] Chin JH, Lee EH, Hwang GS, Hwang JH, Choi WJ. Prediction of Fluid

Prediction of fluid responsiveness using dynamic preload indices in patients undergoing robot-assisted surgery with pneumoperitoneum in the Trendelenburg position. Anaesthesia and Intensive Care 2013;41(4):515-522. DOI: 10.1177/0310057X1304100413

[15] Choi EM, Na S, Choi SH, An J, Rha KH, Oh YJ. Comparison of volume-controlled and pressure-controlled ventilation in steep Trendelenburg position for robot-assisted laparoscopic radical prostatectomy. Journal of Clinical Anesthesia 2011;23(3):183-188. DOI: 10.1016/j.jclinane.2010.08.006

[16] Choi SH, Lee SJ, Rha KH, Shin SK, Oh YJ. The effect of pneumoperitoneum and Trendelenburg position on acute cerebral blood flow-carbon dioxide reactivity under sevoflurane anaesthesia. Anaesthesia 2008;63(12):1314-1318. DOI: 10.1111/j.1365-2044.2008.05636.x

[17] Dal Moro F, Crestani A, Valotto C, Guttilla A, Soncin R, Mangano A, Zattoni F. Anesthesiologic effects of transperitoneal versus extraperitoneal approach during robot-assisted radical prostatectomy: results of a prospective randomized study. International Brazilian Journal of Urology 2015;41(3):466-472. DOI: 10.1590/S1677-5538.IBJU.2014.0199

[18] Doe A, Kumagai M, Tamura Y, Sakai A, Suzuki K. A comparative analysis of the effects of sevoflurane and propofol on cerebral oxygenation during steep Trendelenburg position and pneumoperitoneum for robotic-assisted laparoscopic prostatectomy. Journal of Anesthesia 2016;30(6):949-955. DOI: 10.1007/s00540-016-2241-y

[19] Falabella A, Moore-Jeffries E, Sullivan MJ, Nelson R, Lew M. Cardiac function during steep Trendelenburg position and CO_2 pneumoperitoneum for robotic-assisted prostatectomy: a trans-oesophageal Doppler probe study. The International Journal of Medical Robotics and Computer Assisted Surgery 2007;3(4):312-315. DOI: 10.1002/rcs.165

[20] Haas S, Haese A, Goetz AE, Kubitz JC. Haemodynamics and cardiac function during robotic-assisted laparoscopic prostatectomy in steep Trendelenburg position. The International Journal of Medical Robotics and Computer Assisted Surgery 2011;7(4): 408-413. DOI: 10.1002/rcs.410

[21] Kalmar AF, Foubert L, Hendrickx JF, Mottrie A, Absalom A, Mortier EP, Struys MMRF. Influence of steep Trendelenburg position and CO_2 pneumoperitoneum on cardiovascular, cerebrovascular, and respiratory homeostasis during robotic prostatectomy. British Journal of Anaesthesia 2010;104(4):433-439. DOI: 10.1093/bja/aeq018

[22] Lestar M, Gunnarsson L, Lagerstrand L, Wiklund P, Odeberg-Wernerman S. Hemodynamic perturbations during robot-assisted laparoscopic radical prostatectomy in 45° Trendelenburg position. Anesthesia & Analgesia 2011;113(5):1069-1075. DOI: 10.1213/ANE.0b013e3182075d1f

[23] Oksar M, Akbulut Z, Ocal H, Balbay MD, Kanbak O. Robotic Prostatectomy: The Anesthetist's View for Robotic Urological Surgeries, a Prospective Study. Brazilian Journal of Anesthesiology (English Edition) 2014;64(5):307-313. DOI: 10.1016/j.bjan.2013.10.009

[24] Ono N., Nakahira J., Nakano S., Sawai T, Minami T. Changes in cardiac function and hemodynamics during robot-assisted laparoscopic prostatectomy with steep head-down tilt: a prospective observational study. BMC Research Notes 2017;10(1):341. DOI: 10.1186/s13104-017-2672-z

[25] Rosendal C, Markin S, Hien MD, Motsch J, Roggenbach J. Cardiac and hemodynamic consequences during capnoperitoneum and steep Trendelenburg positioning: lessons learned from robot-assisted laparoscopic prostatectomy. Journal of Clinical Anesthesia 2014;26(5):383-389. DOI: 10.1016/j.jclinane.2014.01.014

[26] Danic MJ, Chow M, Alexander G, Bhandari A, Menon M, Brown M. Anesthesia considerations for robotic-assisted laparoscopic prostatectomy: a review of 1,500 cases. Journal of Robotic Surgery 2007;1(2):119-123. DOI: 10.1007/s11701-007-0024-z

[27] Darlong V, Kunhabdulla NP, Pandey R, Chandralekha Punj J, Garg R, Kumar R. Hemodynamic changes during robotic radical prostatectomy. Saudi Journal of Anaesthesia 2012;6(3):213-218. DOI: 10.4103/1658-354X.101210

[28] Kim NY, Han DW, Koh JC, Rha KH, Hong JH, Park JM, Kim SY. Effect of Dexmedetomidine on Heart Rate-Corrected QT and Tpeak – Tend Intervals During Robot-Assisted Laparoscopic Prostatectomy With Steep Trendelenburg Position: A Prospective, Randomized, Double-Blinded, Controlled Study. Medicine (Baltimore) 2016;95(19):e3645. DOI: 10.1097/MD.0000000000003645

[29] Lutfarakhmanov II, Syrchin EYu, Galeev IR, Mironov PI, Pavlov VN. Changes in Central hemodynamics during robot-assisted radical prostatectomy depending on the type of anesthesia. Russian Journal of Anaesthesiology and Reanimatology (Anesteziologiya i Reanimatologiya). 2020;6:69-76. DOI: 10.17116/anaesthesiology202006169

[30] Matsuoka T, Ishiyama T, Shintani N, Kotoda M, Mitsui K, Matsukawa. Changes of cerebral regional oxygen saturation during pneumoperitoneum and Trendelenburg position under propofol anesthesia: a prospective observational study. BMC Anesthesiology. 2019;19(1):72. DOI: 10.1186/s12871-019-0736-4

[31] Ozdemir M, Bakan N, Sahın OT, Kurtcelebı N, Erbesler ZA, Tunca ST. The Comparison of Sevoflurane-Remifentanyl and Propofol-Remifentanyl in Robotic Prostatectomies. Journal of Clinical and Analytical Medicine 2013;4(4):313-317. DOI: 10.4328/JCAM.1018

[32] Takechi K, Kitamura S, Shimizu I, Yorozuya T. Lower limb perfusion during robotic-assisted laparoscopic radical prostatectomy evaluated by near-infrared spectroscopy: an observational prospective study. BMC Anesthesiology 2018:18;114. DOI: 10.1186/s12871-018-0567-8

[33] Tsvetanova K. The Influence of a Pneumoperitoneum on a Cardio-Vascular System and Central Hemodynamics in the Medical Cases of Robotic and Laparoscopic Gynecological Surgeries. International Journal of Science and Research (IJSR) 2016; 5(4): 968-974. https://www.ijsr.net/search_index_results_paperid.php?id=NOV162728

[34] Meininger D, Westphal K, Bremerich DH, Runkel H, Probst M, Zwissler B, Byhahn C. Effects of posture and prolonged pneumoperitoneum on hemodynamic parameters during laparoscopy. World Journal of Surgery 2008;32(7):1400-1405. DOI: 10.1007/s00268-007-9424-5

[35] Jaju R, Jaju PB, Dubey M, Mohammad S, Bhargava AK. Comparison of volume controlled ventilation and pressure controlled ventilation in patients undergoing robot-assisted pelvic surgeries: An open-label trial. Indian Journal of Anaesthesia. 2017;61(1):17-23. DOI: 10.4103/0019-5049.198406

[36] Matanes E, Weissman A, Rivlin A, Lauterbach R, Amit A, Wiener Z, Lowenstein L. Effects of Pneumoperitoneum and the Steep Trendelenburg Position on Heart Rate Variability and Cerebral Oxygenation during Robotic Sacrocolpopexy. Journal of Minimally Invasive Gynecology 2018;25(1):70-75. DOI: 10.1016/j.jmig.2017.07.009

[37] Kalmar AF, Dewaele F, Foubert L, Hendrickx JF, Heeremans EH, Struys MMRF, Absalom A. Cerebral haemodynamic physiology during steep Trendelenburg position and CO_2 pneumoperitoneum. British Journal of Anaesthesia 2012;108(3):478-484. DOI:10.1093/bja/aer448

[38] La Falce S, Novara G, Gandaglia G, Umari P, De Naeyer G, D'Hondt F, Beresian J, Carette R, Penicka M, Mo Y, Vandenbroucke G, Mottrie A. Low Pressure Robot-assisted Radical Prostatectomy With the AirSeal System at OLV Hospital: Results From a Prospective Study. Clinical Genitourinary Cancer 2017;15(6):e1029-e1037. DOI: 10.1016/j.clgc.2017.05.027

[39] Luo CF, Tsai YF, Chang CH, Wu CT, Yu HP. Increased oxidative stress and gut ischemia caused by prolonged pneumoperitoneum in patients undergoing robot-assisted laparoscopic radical prostatectomy. Acta Anaesthesiologica Taiwanica 2011;49(2):46-49. DOI: 10.1016/j.aat.2011.05.010

[40] Meininger D, Zwissler B, Byhahn C, Probst M, Westphal K, Bremerich DH. Impact of overweight and pneumoperitoneum on hemodynamics and oxygenation during prolonged laparoscopic surgery. World Journal of Surgery 2006;30(4):520-526. DOI: 10.1007/s00268-005-0133-7.

[41] Ozgen SU, Ozveren B, Kilercik M, Aksu U, Ay B, Tufek I, Kural AR, Turkeri LN, Toraman F. Ischemia modified albumin: does it change during pneumoperitoneum in robotic prostatectomies? International Brazilian Journal of Urology 2016;42(1):69-77. DOI: 10.1590/S1677-5538.IBJU.2014.0677

[42] Schramm P, Treiber AH, Berres M, Pestel G, Engelhard K, Werner C, Closhen D. Time course of cerebrovascular autoregulation during extreme Trendelenburg position for robotic-assisted prostatic surgery. Anaesthesia 2014;69(1):58-63. DOI:10.1111/anae.12477

[43] Lasser MS, Renzulli J II, Turini GA III, Haleblian G, Sax HC, Pareek G. An unbiased prospective report of perioperative complications of robot-assisted laparoscopic radical prostatectomy. Journal of Urology 2010;75(5):1083-1089. DOI: 10.1016/j.urology.2009.09.082

[44] Lowenstein L, Mustafa M, Burke YZ, Mustafa S, Segal D, Weissman A. Steep Trendelenburg position during robotic sacrocolpopexy and heart rate variability. European Journal of Obstetrics & Gynecology and Reproductive Biology 2014;178:66-69. DOI: 10.1016/j.ejogrb.2014.03.046

[45] Raimondi F, Colombo R, Costantini E, Marchi A, Corona A, Fossali T, Borghi B, Figini S, Guzzetti S, Porta A. Effects of laparoscopic radical prostatectomy on intraoperative autonomic nervous system control of hemodynamics. Minerva Anestesiologica. 83(12):1265-1273.2017. DOI: 10.23736/S0375-9393.17.12024-9

[46] Lee LC. Cardiopulmonary collapse in the wake of robotic surgery. AANA Journal 2014;82:231-234.

[47] Sharma A, Berkeley A. Intraoperative drug-eluting stent thrombosis in a patient undergoing robotic prostatectomy. Journal of

Clinical Anesthesia 2009;21(7):517-520. DOI: 10.1016/j.jclinane.2008.11.013

[48] Thompson J. Myocardial infarction and subsequent death in a patient undergoing robotic prostatectomy. AANA Journal 2009;77(5):365-371.

[49] Yamabe F, Mitsui Y, Hoshino O, Shimizu T, Kasahara M, Kobayashi H, Nakajima K. Temporary pacemaker insertion for severe bradycardia following pneumoperitoneum during robot-assisted radical prostatectomy: a case report. BMC Surgery 2020;20(1):238. doi: 10.1186/s12893-020-00902-9

[50] Gainsburg DM. Anesthetic concerns for robotic-assisted laparoscopic radical prostatectomy. Minerva Anestesiologica 2012;78(5):596-604.

[51] Lauridsen SV, Tønnesen H, Jensen BT, Neuner B, Thind P, Thomsen T. Complications and health-related quality of life after robot-assisted versus open radical cystectomy: a systematic review and Meta-analysis of four RCTs. Systematic Reviews 2017;6(1):150. DOI: 10.1186/s13643-017-0547-y

[52] Roviello F, Piagnerelli R, Ferrara F, Scheiterle M, De Franco L, Marrelli D. Robotic single docking total colectomy for ulcerative colitis: First experience with a novel technique. International Journal of Surgery 2015;21:63-67. DOI: 10.1016/j.ijsu.2015.07.642

[53] Seamon LG, Bryant SA, Rheaume PS, Kimball KJ, Huh WK, Fowler JM, Phillips GS, Cohn DE. Comprehensive Surgical Staging for Endometrial Cancer in Obese Patients. Obstetrics & Gynecology 2009;114(1):16-21. DOI: 10.1097/AOG.0b013e3181aa96c7

[54] Al-Mazrou AM, Chiuzan C, Kiran RP. The robotic approach significantly reduces length of stay after colectomy: a propensity score-matched analysis. International Journal of Colorectal Disease 2017;32(10):1415-1421. DOI: 10.1007/s00384-017-2845-1

[55] Al-Mazrou AM, Baser O, Kiran RP. Propensity Score-Matched Analysis of Clinical and Financial Outcomes After Robotic and Laparoscopic Colorectal Resection. Journal of Gastrointestinal Surgery 2018;22:1043-1051. DOI: 10.1007/s11605-018-3699-8

[56] Al-Temimi MH, Chandrasekaran B, Agapian J, Peters WR Jr, Wells KO. Robotic versus laparoscopic elective colectomy for left side diverticulitis: a propensity score-matched analysis of the NSQIP database. International Journal of Colorectal Disease 2019;34(8):1385-1392. DOI: 10.1007/s00384-019-03334-x

[57] Benlice C, Aytac E, Costedio M, Kessler H, Abbas MA, Remzi FH, Gorgun E. Robotic, laparoscopic, and open colectomy: a case-matched comparison from the ACS-NSQIP. International Journal of Medical Robotics and Computer Assisted Surgery. 2017;13(3). DOI: 10.1002/rcs.1783

[58] Chalmers D, Cusano A, Haddock P, Staff I, Wagner J. Are Preexisting Retinal and Central Nervous System-Related Comorbidities Risk Factors for Complications Following Robotic-Assisted Laparoscopic Prostatectomy? International Brazilian Journal of Urology 2015;41(4):661-668. DOI: 10.1590/S1677-5538.IBJU.2014.0464

[59] Feinberg AE, Elnahas A, Bashir S, Cleghorn MC, Quereshy FA. Comparison of robotic and laparoscopic colorectal resections with respect to 30-day perioperative morbidity. Canadian Journal of Surgery 2016;59(4):262-267. DOI: 10.1503/cjs.016615

[60] Han H, Cao Z, Qin Y, Wei X, Ruan Y, Cao Y, He J. Morbid obesity is

adversely associated with perioperative outcomes in patients undergoing robot-assisted laparoscopic radical prostatectomy. Canadian Urological Association Journal 2020;14(11):574-581. DOI: 10.5489/cuaj.6389

[61] Hellan M, Anderson C, Ellenhorn JD, Paz B, Pigazzi A. Short-term outcomes after robotic-assisted total mesorectal excision for rectal cancer. Annals of Surgical Oncology 2007;14(11):3168-3173. DOI: 10.1245/s10434-007-9544-z

[62] Paley PJ, Veljovich DS, Shah CA, Everett EN, Bondurant AE, Drescher CW, Peters WA 3rd. Surgical outcomes in gynecologic oncology in the era of robotics: analysis of first 1000 cases. American Journal of Obstetrics & Gynecology. 2011; 204(6):551.e1-551.e9. DOI: 10.1016/j.ajog.2011.01.059

[63] Piegeler T, Dreessen P, Graber SM, Haile SR, Schmid DM, Beck-Schimmer B. Impact of intraoperative fluid administration on outcome in patients undergoing robotic-assisted laparoscopic prostatectomy – a retrospective analysis. BMC Anesthesiology 2014;14:61. DOI: 10.1186/1471-2253-14-61

[64] Trinh QD, Sammon J, Sun M, Ravi P, Ghani KR, Bianchi M, Jeong W, Shariat SF, Hansen J, Schmitges J, Jeldres C, Rogers CG, Peabody JO, Montorsi F, Menon M, Karakiewicz PI. Perioperative outcomes of robot-assisted radical prostatectomy compared with open radical prostatectomy: results from the nationwide inpatient sample. European Urology 2012;61(4):679-985. DOI: 10.1016/j.eururo.2011.12.027

[65] Sakai Y, Yasuo MT, Oyama T, Murakami C, Kakuta N, Tanaka K. Noninvasive continuous blood pressure monitoring by the ClearSight system during robot-assisted laparoscopic radical prostatectomy. The Journal of Medical Investigation. 2018;65(1.2): 69-73. DOI: 10.2152/jmi.65.69

[66] Aceto P, Beretta L, Cariello C, Claroni C, Esposito C, Forastiere EM, Guarracino F, Perucca R, Romagnoli S, Sollazzi L, Cela V, Ercoli A, Scambia G, Vizza E, Ludovico GM, Sacco E, Vespasiani G, Scudeller L, Corcione A on behalf of Società Italiana di Anestesia Analgesia Rianimazione e Terapia Intensiva (SIAARTI), Società Italiana di Ginecologia e Ostetricia (SIGO), and Società Italiana di Urologia (SIU). Joint consensus on anesthesia in urologic and gynecologic robotic surgery: specific issues in management from a task force of the SIAARTI, SIGO, and SIU. Minerva Anestesiologica. 2019;85(8):871-885.

Chapter 5

Hemodynamic Alterations in Multiple Sclerosis

Aise Seda Artis

Abstract

Multiple Sclerosis is an autoimmune disease of the central nervous system. It is a demyelinating and neurodegenerative condition, however, changes in the vasculature can occur and play a role in the pathophysiology. Cardiac and vascular risk factors contribute to the disease severity. Understanding the occurring hemodynamic changes may potentially lead to improved diagnosis, better patient management, and prevention of disease progression. This paper discusses the hemodynamic impairment in multiple sclerosis focusing on both the cerebral and cervical regions and presents an up-to-date review of the literature.

Keywords: multiple sclerosis, cerebral hemodynamics, cervical hemodynamics, perfusion, oxidant damage

1. Introduction

Multiple Sclerosis (MS) is a chronic disease of the central nervous system (CNS), today affecting approximately 2.8 million people worldwide (35.9 per 100,000 population) [1]. It leads to demyelination and diffuse neurodegeneration in both brain and spinal cord gray matter (GM) and white matter (WM) of the brain and spinal cord [2].

The natural history of the disease seems to be divided into 2 distinct phases for most of the patients with MS (PwMS): 1) initially most of them present recurring clinical symptoms followed by total or partial recovery, this form is called relapsing-remitting MS (RRMS); 2) after 10–15 years, the pattern becomes progressive in up to 50% of untreated patients. This second phase is defined as secondary progressive MS (SPMS) and determined by a slow but steady progression in neurologic deficit associated with CNS degeneration. Alternatively, 15% of PwMS can present with a progressive form at the onset, a progressive clinical decline without superimposed exacerbations, and this form is called primary progressive MS (PPMS) [3]. There is also a pre-clinical stage in which a combination of genetic and environmental factors triggers the disease [2].

MS is an autoinflammatory condition. However, changes in the vasculature can occur and contribute to pathophysiology [4]. The pathophysiology of hemodynamic impairment in MS is multifactorial and, at least partly, secondary to the downstream effects of the neuro-inflammatory cascades [2]. The incidence of vascular comorbidities was previously reported as up to 50% but continues rising in the MS population, making it important to understand their impacts on outcomes [5, 6].

Cardiovascular risk factors are known to contribute to MS disease severity. Smoking is associated with a higher lesion burden and more severe brain atrophy in people with MS (PwMS). This relationship becomes even more prominent if there are multiple cardiovascular comorbidities [2, 7]. Comorbidity is associated with greater diagnostic delays, worse magnetic resonance imaging (MRI) outcomes, increased disability at diagnosis, and increased risk of disease progression in MS [8, 9]. There is a direct relationship between cardiovascular risk factors and clinical status as measured by the Expanded Disability Status Scale (EDSS) [10]. A large cohort study reported that hypertension and heart disease were associated with brain atrophy and obesity was associated with lesion volumes [8]. Also, high sodium intake, a regulating factor of blood pressure, seems to be linked to MS disease activity [11]. Understanding the hemodynamic changes in MS could potentially lead to better management of patients and improved diagnosis and prevention of disease progression.

2. Cervical hemodynamic changes in MS

When compared with the healthy population, PwMS show different patterns of vascular morphology in the neck, with respect to aging. The arteries supplying the CNS are possibly subject to particular atherosclerotic harm in MS [12]. The vascular cross-sectional area (CSA) in the neck is crucial to further understanding the associations between the extracranial and intracranial vascular changes.

Ranadive et al. reported significantly altered arterial function, as shown by decreased carotid artery compliance, but not structure in PwMS compared with the control matched for age, sex, height, and weight [12]. However, recent data suggests a smaller arterial cross-sectional area of the main and secondary arteries (common, internal, and external carotid arteries and vertebral artery, respectively) in PwMS. The CSA of the carotid and vertebral arteries is reduced [7]. Besides, significantly higher carotid intima-media thickness was reported in PwMS without cardiovascular disease compared to the healthy group, suggesting that PwMS have a predisposition to atherosclerosis [13]. In MS, even without the presence of cardiovascular disease carotid and vertebral arterial CSA was reduced [7]. A 5-year follow-up study checked the neck vessel CSA in PwMS and healthy controls during 5 years by 3 Tesla (3 T) MRI using 2-dimensional (2D) neck MRI angiography. At baseline, they observed no difference in CSA between the groups. The monitoring revealed a decrement in CSA of the common carotid artery – internal carotid artery, vertebral artery, and internal jugular vein (IJV), regardless of the disease phenotype. Interestingly, PwMS without the cardiovascular disease had significantly greater change than PwMS with cardiovascular disease for IJVs at all levels. Their observation of longitudinally changing IJV CSA may suggest a potential link between IJV CSA and the disease course in MS [14]. Heterogeneity of PwMS groups among the studies may contribute to explaining the difference in the results. Still, some hypertension-perfusion interactions might be considered in PwMS without cardiovascular disease when the lately lowered hypertension threshold to >130/80 mmHg in the guidelines taken into consideration.

Some cross-sectional studies revealed a greater prevalence of morphologic and hemodynamic alterations of extracranial venous drainage pathways in PwMS [15, 16]. Zamboni asserted that narrowing of the veins in the neck could be causing iron accumulation in the brain and spinal cord, triggering the inflammatory autoimmune response and proposed his hypothesis of Chronic Cerebrospinal Venous Insufficiency (CCSVI) as a cause of MS [15, 17]. The distinct abnormalities observed in the intracranial and extracranial veins in PwMS defined by Zamboni in 2009. Zamboni criteria include: 1) Reflux in the IJVs in sitting and supine posture; 2) Reflux in the deep cerebral veins (DCV); 3) High-resolution B-mode evidence of IJV stenoses; 4) Flow not doppler-detectable in the IJVs. 5) Reverted postural control of the main cerebral venous outflow pathways [15].

CCSVI is only seen in MS patients and not in other neurodegenerative disorders or healthy individuals [15, 18]. It causes venous hypertension in the dural sinuses which can alter intracranial compliance. Subsequently, it may alter the CSF dynamics, affecting the ability of the vasculature to dissipate arterial pulsatility and provide smooth capillary blood flow that is known as the windkessel mechanism [19]. CCSVI is characterized by multiple areas of stenosis of the extracranial venous draining pathways, namely the IJVs and the azygous veins, with collateral formation. Normally the blood leaves the brain by postural and respiratory mechanisms, where the venous outflow increases during inspiration. The normal venous drainage pathway of the blood leaving the brain is via the IJV and the azygous vein (in the supine posture) and the vertebral vein (in the upright position). And for the spinal cord, the main route for venous drainage is the azygous vein [15, 18]. CCSVI can be easily assessed using doppler sonography or magnetic resonance venography.

Compared with healthy controls, PwMS exhibit reduced venous flow in IJV [16]. Haacke et al. confirmed the venous flow abnormalities in all clinical forms of MS. In their study, IJV stenosis was more prevalent in PwMS, and IJV carried significantly less flow when compared to the nonstenotic group [20]. PwMS show a higher frequency of secondary neck veins and larger cross-sectional areas when adjusted for all cardiovascular factors (including body mass index, hypertension, heart disease, smoking history, and age) [7]. Accordingly, flow in the paraspinal venous collaterals is higher in PwMS and exacerbated by venous stenosis. Collateral drainage may be a compensatory response to IJV flow reduction [16]. Both CCSVI and small IJVs (with a cross-sectional area of less than 0.4 cm^2) seem to influence or follow MS severity. However, only small IJVs are an independent factor. Thus, small IJVs with restricted outflow, which might be aspects of CCSVI different from the criteria originally described by Zamboni, emerge as a cofactor in the multifactorial pathophysiology of MS [21]. Considering the morphology of neck veins demographic factors may also be confounding between PwMS and healthy groups [7]. Notably, no significant difference in PwMS regarding IJV CSA and flow rates has also been reported [22, 23].

With CCSVI, Dr. Zamboni also proposed venous percutaneous transluminal angioplasty (so-called "liberation treatment" or "liberation therapy") as a potential therapy for PwMS [24]. Today liberation treatment is a controversial treatment. CCSVI-like venous anomalies seem unlikely to affect cerebral blood flow (CBF) in PwMS according to some researchers [19]. Recent studies do not support the continued use of angioplasty for the extracranial jugular and/or azygous venous narrowing to improve patient-reported outcomes, chronic MS symptoms, or the disease course of MS [25–27]. But Zamboni et al. reported that venoplasty decreases new cerebral lesions at 1 year in RRMS and SPMS after the angioplasty [28]. It seems that the debate will continue until more detailed and comprehensive long-term studies are provided. Future studies should investigate CSA of the arterial and venous systems of neck vessels in more detail at disease onset, when the presence of cardiovascular risks is minimal, with a comprehensive approach.

3. Cerebral hemodynamic changes in MS

Changes in brain vasculature contribute to the pathophysiology of MS [4]. Especially the periventricular veins are vulnerable to ischemia and plaque formation due to their hydrodynamic properties. Demyelination and lesion formation is associated with the breakdown of the blood–brain barrier around postcapillary venules, where MS lesions are commonly located [29]. MS is characterized by changes in the WM in the periventricular region and is also associated with enlarged lateral ventricles. The brain atrophy seen in MS might be primarily responsible for ventricular enlargement [19, 30]. The cerebral venous system plays an important role in the intracranial hemodynamic/cerebrospinal fluid regulatory system. This influence both the perfusion of the brain parenchyma and the dynamics of the CSF system. The generally accepted opinion is that cerebral perfusion is globally reduced in MS [19].

Cerebral perfusion is usually measured as CBF. It represents the blood volume that passes through a given volume of brain parenchyma per time unit [31]. Absolute measurements of cerebral blood volume (CBV), CBF, and mean transit time (MTT) reflect the overall perfusion of chronic lesions in PwMS. Long ago there were reports of reduced cerebral perfusion of the WM and GM of PwMS, which received little attention at the time [31–34].

Vasoreactivity reflects the ability of microvasculature to adapt to a changing microenvironment. A dynamic process called cerebral vasoregulation redistributes CBF depending on the fluctuating metabolic demands such as oxygen and glucose delivery and blood pressure variations. The neurovascular unit (NVU) broadly describes the relationship between brain cells and their blood vessels, regulating local, regional, and global perfusions. The other related terms are neurovascular coupling, cerebrovascular reactivity, and hemodynamic response function [35–37]. Both cellular and extracellular components are involved in the regulatory function of the NVU [35]. The cellular components are the neurons, perivascular astrocytes, microglia, pericytes, endothelial cells, and the basement membrane. Glial cell intermediaries facilitate the ability of neurons to adequately convey metabolic needs to cerebral vasculature for sufficient oxygen and nutrient perfusion [36]. The NVU is responsible for the maintenance of a highly selective blood-brain barrier and cerebral homeostasis, as well as the control of CBF through the cerebral metabolic rate of oxygen consumption [38]. It facilitates the relationship among neuronal activity, hemodynamic factors, and cell-to-cell signaling. Suboptimal blood delivery during neuronal activities caused by disrupted NVU coupling may eventually lead to neuronal dysfunction and degeneration in a chronic state.

The most characteristic brain tissue injury in MS is primary demyelination, with partial preservation of axons. Neural inflammation causes neurodegeneration together with demyelination; both of which are also worsened by tissue hypoxia. Inflammation further contributes to tissue hypoxia through impaired CBF and hypoperfusion which are the result of NVU dysfunction [4].

MRI is an important diagnostic tool for MS because it produces images of lesions in the brain and spinal cord. Widespread microglial activation observed in MS in areas surrounding the focal lesions is called normal-appearing WM (NAWM) [39]. Also, if on conventional T2-weighted (T2w) MRI normal-looking GM is histopathologically abnormal then this is referred to as normal-appearing GM (NAGM) [40].

Investigated absolute measures of flow and volume revealed decreased CBF in the NAWM of patients with RRMS [41]. Interestingly, another group observed elevated CBF and CBV in NAWM of RRMS patients several weeks before focal leakage of the blood–brain barrier and plaque formation [42]. In NAWM, hypoperfusion has been associated with persistent low-grade inflammation, metabolic or vascular dysfunction, or primary ischemia [43, 44]. Conversely, increased perfusion preceding focal WM lesion formation could indicate an increased inflammatory response before tissue damage in NAWM [45]. Many studies suggest that CBF is decreased in several regions of NAWM and NAGM compared to the healthy population [42, 43, 46–55]. One study reported lower CBF and higher MTT, consistent with reduced perfusion, in WM lesions compared to NAWM in patients with early MS [44]. However another one observed increased CBV and CBF in RRMS, in patients with high inflammatory lesion load, underlining the role of global modified microcirculation prior to leakage of the blood-brain barrier in the pathophysiology of MS [56]. The perfusion alterations in RRMS seem to be independent of GM volume atrophy, which presents in all types of MS [57]. Hemodynamic changes in both WM and GM seem to occur even at the earliest stages of MS. Interestingly, a greater reduction of NAWM CBF was found in PPMS compared to RRMS [47, 49]. This finding is important because people with RRMS tend to have more brain lesions with more inflammatory cells.

Clinically isolated syndrome (CIS) refers to a single clinical attack of CNS inflammatory demyelinating symptoms that are suggestive of MS [58]. Reduced cerebral perfusion has also been observed in the NAWM of patients with CIS [50]. This is not surprising in CIS, since those patients already show evidence of NAWM [59]. But the decrement in CIS is at a lesser degree than in RRMS [50]. One of the earlier studies using DSC MRI demonstrated that acute gadolinium-enhancing lesions in PwMS have higher relative CBV when compared with their own contralateral NAWM [60]. However, possibly it is inappropriate to use NAWM as a reference by itself since it is also a pathological state [50]. On the other hand, CBF in the subcortical NAGM was not reduced in patients with CIS when compared to the healthy group [50]. Nevertheless, there is an observation of altered perfusion in the deep GM in both RRMS and CIS: In CIS all measured NAWM and deep NAGM regions had significantly higher CBV and MTT values, while averaged deep GM regions had significantly lower CBF values. In RRMS, the same regions showed lower CBF values than those of healthy volunteers and even lower CBV and CBF values compared to patients with CIS [61]. Hemodynamic changes in CIS present differently than the other forms of MS.

DSC-MRI and arterial spin-labeling studies of NAWM, cerebral cortex, subcortical GM, and deep GM point to a widespread decrease in perfusion in RRMS and progressive MS patients compared to healthy individuals [41, 49, 50, 61, 62]. The findings of a recent transcranial Doppler ultrasonography study also are consistent with the accumulating evidence of decreased cerebral perfusion in MS [4]. These results suggest that cerebral hypoperfusion, regardless of the clinical type, is an early and integral part of MS pathology [63]. The studies also suggest a continuum of reduction in tissue perfusion, beginning in the WM and spreading to GM with the disease progression [50].

4. Possible mechanisms of altered perfusion

Determining whether the cerebral hypoperfusion has primary or secondary etiology is critical in MS. For this purpose, researchers try to investigate every possible mechanism. **Table 1** summarizes the possible causes of altered perfusion in MS that are discussed briefly below.

Autonomic dysfunction resulting from demyelination.
Obliterating perivascular MS lesions.
Chronic cerebrospinal venous insufficiency (CCSVI).
Axonal injury with microglia activation caused by global inflammation.
Oxidative damage due to released toxic inflammatory mediators.

Table 1.
Possible causes of altered perfusion in MS.

Demyelination in MS may cause damage to the autonomic nervous system [64]. Autonomic receptors, both sympathetic and parasympathetic, have a significant role in dynamic cerebral autoregulation. Autonomic cardiac dysfunction is seen in up to 63% of PwMS [65, 66]. It can be present with or without symptoms and may be associated with the presence of brainstem lesions [67]. The symptoms are caused by cerebral hypoperfusion. They are typically induced by standing and quickly resolve when lying flat. The most likely cause is the central dysregulation of sympathetic and parasympathetic outflow to the cardiovascular system [68].

Another possibility would be reduced blood flow associated with obliterating perivascular MS lesions. However, as focal CBF decrease is not in a patchy pattern in MS, this idea can easily be ruled out. Microvessel thrombosis and other structural abnormalities have been observed very exceptionally within MS plaques [69]. Besides the increased CBF in active inflammatory lesions also argue against this theory [63]. Notably, compensatory functional adaptations might also account for MS-related changes in brain perfusion and activity [70].

Zamboni et al. reported a strong relationship between CCSVI and MS [15, 17, 18, 28, 71]. However, Beggs et al. propose that CCSVI-like venous anomalies seem unlikely to account for reduced CBF in PwMS [19]. The results of a phase-contrast MRI study do not support the CCSVI hypothesis that CSF flow decreases in MS patients [30]. Their results favor Beggs et al. who observed increased CSF pulsatility in the aqueduct of Sylvius, which they explain by the mechanisms increasing the hydraulic resistance of the cerebral vascular bed [19]. The results of the studies discussing angioplasty as a treatment of MS also seems in favor of the idea that CCSVI is less likely to take part in the pathophysiology of MS [25, 26].

Astrocytes actively control the blood-brain barrier and regulate CBF. In progressive MS lesions, diffuse pathology is also present in NAWM and NAGM, reflected by diffuse axonal injury with profound microglia activation within a background of a global inflammation of the entire brain and the meninges [72]. The relation between perfusion-weighted and diffusion tensor MRI features in the normal-appearing corpus callosum of patients with RRMS revealed a decreased CBF, which positively correlated with mean diffusivity but not with fractional anisotropy. This observation favors primary ischemia, rather than hypoperfusion secondary to axonal degeneration [51]. Decreased levels of N-acetylaspartate (NAA) indicate reduced axonal metabolism [73]. A positive correlation of cerebral perfusion with NAA levels was present for healthy controls, not for PwMS. The perfusion reduction was greater than would be expected from decreased axonal metabolism or axonal loss alone in MS [74]. Additionally, the excitability of primary motor cortex neurons is increased in progressive MS, which potentially escalates their metabolic demand [75]. So, reduced CBF does not seem to be secondary to axonal degeneration with reduced metabolic demands [76]. On the other hand, a hypothesis proposes that hypoperfusion is the result of neuronal dysfunction, which is the result of oxidative injury in cortical neurons or retrograde neurodegeneration due to axonal injury from demyelination [77]. There is another conflict is on the NVU coupling. Most investigators report diffusely impaired NVU coupling in PwMS [37]. However, some studies such as recent magnetoencephalography (MEG)- fMRI study suggest that NVU coupling is preserved in MS patients [78].

Accumulating evidence proposes that toxic inflammatory mediators and resulting oxidative damage play a prominent role in the pathophysiology of altered cerebral hemodynamics in MS. Iron accumulation in the extracellular space and its uptake into cells within the lesions might increase the susceptibility of the surrounding tissue to free-radical driven demyelination and neurodegeneration, which is likely to be more pronounced in progressive MS [39]. Mitochondrial damage in MS lesions could be mediated by reactive oxygen and nitric oxide (NO) species [79]. It is evident that WM lesions represent the regions, where most current, past, and repetitive inflammatory activities occur that are associated with overproduced NO [37]. The overproduction and prolonged exposure to NO can affect the elasticity of the blood vessels and cause vascular habituation that leads to impaired perfusion, and cerebrovascular response deficit, and neurodegeneration including GM atrophy [37]. Endothelin-1 (ET-1) is a vasoconstrictor secreted by endothelial cells, which acts as the natural counterpart of the vasodilator NO. ET-1 levels elevate in both peripheral blood and cerebrospinal fluid of PwMS [80]. Reactive astrocytes in MS plaques release ET-1 in the cerebral circulation, which participates CBF reduction in MS by inducing arteriolar vasoconstriction [63]. ET-1 upregulation has been associated with reduced extra-ocular blood flow velocities [81]. Besides retinal oxygen metabolism is reportedly affected in MS by increased venular oxygen saturation and lower AV difference [82]. Oxidative stress causes calcium influx into the cytoplasm from the extracellular environment and endoplasmic reticulum or sarcoplasmic reticulum. A rise of calcium levels within astrocytes induces constriction of blood vessels and consequently reduces CBF [83]. Another possible role player is transcription factor hypoxia-inducible factor-1α (HIF-1α). It mediates adaptive responses to oxidative stress by nuclear translocation and regulation of gene expression. Expression of hypoxia-inducible factor-1α and its downstream genes is also elevated in MS [69, 81]. Finally, glutathione levels are reduced in CNS of PwMS [84]. Glutathione serves as an antioxidant that aids in protecting neurons against oxidative damage. Glutathione levels are low also in the periphery [85].

5. Methods for evaluation of perfusion

Reduced cerebral perfusion in both the WM and GM of PwMS is known by single-photon emission computed tomography (SPECT) and positron emission tomography (PET) studies since the 1980s [31–34]. However, the main problem with these studies was a low spatial resolution, and eventually they received little attention at the time [63]. The topic of cerebral perfusion in MS regained interest with the development of more accurate imaging and processing techniques, allowing better visualization and differentiation between WM plaques, NAWM and NAGM. Today commonly Doppler and MRI-based techniques are used and provide better imaging [15, 17, 18, 25, 26, 86].

In the 2000s Doppler ultrasound and invasive selective venography studies helped to coin the term CCSVI [15, 17]. Today the latter is mostly replaced by magnetic resonance venography [86].

Transcranial Doppler sonography is for real-time cerebral vasomotor reactivity assessment. This technique is safe, low-cost, practical, and easy to perform in clinical practice. Lattanzi et al. measured NVU response to hypercapnia by the breath-holding index through this method. The main findings were a decrease in NVU coupling for both RRMS and SPMS groups, and greater impairment in cerebral hemodynamics in SPMS than RRMS, as suggested by lower breath-holding index in SPMS patients [4].

Arterial spin labeling is a non-invasive perfusion-weighted MRI method and uses magnetically labeled arterial blood as the endogenous tracer [87]. It characterizes oxidative stress and provides a completely non-invasive means to measure quantitative CBF within the brain [87]. It is effective in detecting decreased perfusion in GM [46, 48, 55]. A study using arterial spin labeling found an increased WM CBF in both RRMS and SPMS groups [48]. However, the authors did not distinguish NAWM from focal WM lesions [63].

Functional MRI (fMRI) measures brain activity by detecting changes associated with blood flow. This technique relies on the fact that CBF and neuronal activation are coupled, so helps to assess the NVU function. The primary form of fMRI uses the blood-oxygen-level-dependent (BOLD) contrast, which is the MRI contrast of blood deoxyhemoglobin. Turner et al. used fMRI to quantify the extent to which WM affects NVU coupling and cognitive performance [36]. NVU function is generally considered to be a proxy for underlying neural activity originating in GM. The researchers modeled it from multiple brain regions during multiple cognitive tasks. Their results support the idea that intact neural-glial-vascular communication underlies optimal neural and cognitive functioning [36].

Functional-near infrared spectroscopy (fNIRS) measures blood oxygenation changes similar to fMRI. The results of an fNIRS study in the prefrontal cortex indicate that measuring the slope coefficient of oxy- and deoxy-hemoglobin concentrations during walking is reliable for most of the included areas in PwMS [88].

Additionally, dynamic susceptibility contrast (DSC)-MRI, one of the most frequently used techniques for MRI perfusion, can be used to quantitatively assess cerebral perfusion. It can be diagnostic for the acute inflammatory phase of lesion development [89, 90]. DSC-MRI relies on the susceptibility induced signal loss on T2w sequences which results from a bolus of gadolinium-based contrast passing through a capillary bed. The most commonly calculated parameters are CBV, CBF, and MTT [91]. One research reported an increase, and another one reported a decrease in global perfusion in RRMS patients, both using DSC-MRI [44, 56]. The methodological differences in both studies seem to be a cause of this contradiction: The first researchers included newly diagnosed PwMS and used a 1.5 Tesla (T) MRI [44]. On the other hand, the second researchers divided the participants into high- and low-inflammatory groups according to the number of new contrast-enhancing lesions; and had a 3 T MRI [56]. This reflects the intricate spatiotemporal dynamics and heterogeneous disease progression that is characteristic of RRMS [92]. Of course, better imaging provided by 3 T MRI should not be ignored. Another important point is that gadolinium-enhancing areas (MS lesions) show increased CBF on DSC-MRI [42, 43, 60]. So total WM CBF may thus be overestimated when focal

Principles	**Methods**
X-ray based techniques	Selective venography
Nuclear imaging techniques	SPECT
	PET
USG based techniques	Doppler USG
MRI based techniques	fNIRS
	BOLD – fMRI
	MR venography
	DSC-MRI ± SAGE
	Arterial spin labeling
	Whole-brain OEF mapping
Infrared light absorption	Retinal oximetry

**Single-photon emission computed tomography (SPECT), positron emission tomography (PET), ultrasound (USG), magnetic resonance imaging (MRI), blood-oxygen-level-dependent (BOLD), functional MRI (fMRI), functional-near infrared spectroscopy (fNIRS), dynamic susceptibility contrast (DSC)-MRI, spin and gradient-echo (SAGE), oxygen extraction fraction (OEF).*

Table 2.
Principles and methods for evaluation of perfusion in MS.*

lesions were not properly excluded from perfusion measurements [63]. Recently, an advanced DSC-MRI method was used to quantitatively measure global and capillary-sized CBF, CBV, and MTT in RRMS: a combined spin- and gradient-echo (SAGE) perfusion imaging method. It showed that compared to NAWM, lesion regions-of-interest (ROIs) had significantly reduced perfusion (CBF and CBV) and increased MTT, as well as reduced WM microstructural integrity. The changes within lesion ROIs were associated with altered WM microstructural integrity. WM microstructural integrity displayed weak positive correlations in lesion ROIs with perfusion parameters [92].

Additionally, whole-brain oxygen extraction fraction mapping for measuring lesion-specific and regional oxygen extraction fraction abnormalities may serve as a useful quantitative marker of tissue oxygen utilization in MS [93].

Besides, by applying optical principles similar to those used in pulse oximetry, retinal oximetry reflects the retinal oxygen metabolism changes. And it has the potential of becoming a new biomarker in MS [82]. Above mentioned methods are listed in **Table 2**.

6. Clinical implications

Cerebrovascular hemodynamic insufficiency in MS may have clinical implications due to its contributions to MS symptomatology. Reduced CBF may contribute to focal lesion formation [63]. West et al. observed a lower cerebral metabolic rate of oxygen ($CMRO_2$) in PwMS compared to healthy volunteers. After controlling for demographic and disease characteristics (i.e., age, education, disability, lesion volume), $CMRO_2$ predicted increased fatigue and reduced cognitive performance in MS patients. PwMS with higher $CMRO_2$ have a reduced fractional anisotropy, a useful measure of connectivity in the brain, in NAWM [94]. When metabolic demand is increased by the activation of cerebral areas, as during cognitive tasks, blood supply may not suffice due to the reduced perfusion reserve [95]. Impaired NVU coupling increases the hazard of cerebral ischemic events. Epidemiological data showed that PwMS are at higher risk of stroke [76].

CBF in NAWM correlates with clinical disability [49]. But in GM CBF correlates with neuropsychological dysfunctions [47]. Beyond global CVR deficits and neurodegeneration found in MS, the integrity of specific functional networks may be more affected than others [37]. The evaluation of cerebral hemodynamic status may stratify individual vascular risk [4].

In the light of pathophysiological studies, clinicians try some approaches with the efforts to minimize the disease progression. There are dietary approaches like Wahls diet [96]. To overcome the oxidative stress altering hemodynamics the importance of dietary nitrate intake should not be underestimated. Dietary nitrate, derived in the diet primarily from vegetables, is converted to NO in the body. There are also therapeutical intervention trials. In order to restore CBF in RRMS, Hostenbach et al. administered ET-1 antagonist bosentan. In the study, the results showed that CBF in the patients was not different from that of the healthy controls and bosentan did not increase CBF. The authors commented that it has no effect when CBF values are within the normal range [53]. Similarly, Shahrampour et al. administered N-acetylcysteine (NAC) to RRMS and PPMS patients. Interestingly, certain brain regions experienced an increase while others experienced a decrease in CBF following the treatment. This highlights the notion that NAC does not have a uniform effect on the brain but appears to target specific regions that are affected in MS. NAC administration was associated with altered resting CBF and qualitative improvements in cognition and attention in PwMS [97].

7. Conclusion

The pathways of tissue damage in MS are heterogeneous and not completely understood. The studies exploring the relationships between cerebral hemodynamics, functional impairment, disease course, and therapeutic response may reasonably allow to improve the understanding of MS pathophysiology and translate in implications for clinical practice [4]. The NVU dysfunction and interplay between inflammatory and vascular changes seem to be the key players in the pathophysiology of MS. Altered cervical and cerebral perfusion in MS is associated with reduced brain integrity. WM and GM integrity changes lead to a higher risk of relapses, disability, and disease-modifying therapy escalation. So, understanding the hemodynamic effects is very critical to help PwMS.

There are debatable issues for cervical and cerebral hemodynamics in MS. These disagreements can at least be partially explained by the heterogeneity of inclusion criteria and methods. The variety of the tools, techniques, and used protocols cause different outcomes.

In order to avoid additional factors damaging the brain, to provide improved diagnosis, superior patient management and prevention of disease progression, to define reliable biomarkers, and to design novel therapeutic strategies, a thorough understanding of the hemodynamic changes in MS is critical. Future research especially follow-up studies with larger populations under different activity conditions would ease answering today's questions.

Conflict of interest

The author declares no conflict of interest.

Author details

Aise Seda Artis[1,2]

1 Istanbul Medeniyet University, Istanbul, Turkey

2 Vistula University, Warsaw, Poland

*Address all correspondence to: aseda@yahoo.com

References

[1] Walton C, King R, Rechtman L, Kaye W, Leray E, Marrie RA, et al. Rising prevalence of multiple sclerosis worldwide: Insights from the Atlas of MS, third edition. Multiple Sclerosis Journal. 2020;**26**(14):1816-1821

[2] Baecher-Allan C, Kaskow BJ, Weiner HL. Multiple sclerosis: Mechanisms and immunotherapy. Neuron. 2018;**97**(4):742-768

[3] Lublin FD, Reingold SC. Defining the clinical course of multiple sclerosis: Results of an international survey. National Multiple Sclerosis Society (USA) Advisory Committee on Clinical Trials of New Agents in Multiple Sclerosis. Neurology. 1996;**46**(4):907-911

[4] Lattanzi S, Acciarri MC, Danni M, Taffi R, Cerqua R, Rocchi C, et al. Cerebral hemodynamics in patients with multiple sclerosis. Multiple Sclerosis and Related Disorders. 2020;**44**:102309

[5] Dossi DE, Chaves H, Heck ES, Rodriguez Murúa S, Ventrice F, Bakshi R, et al. Effects of systolic blood pressure on brain integrity in multiple sclerosis. Frontiers in Neurology. 2018;**9**:487

[6] Marrie RA, Patel R, Figley CR, Kornelsen J, Bolton JM, Graff L, et al. Diabetes and anxiety adversely affect cognition in multiple sclerosis. Multiple Sclerosis and Related Disorder. 2019; **27**:164-170

[7] Belov P, Jakimovski D, Krawiecki J, Magnano C, Hagemeier J, Pelizzari L, et al. Lower arterial cross-sectional area of carotid and vertebral arteries and higher frequency of secondary neck vessels are associated with multiple sclerosis. American Journal of Neuroradiology. 2018;**39**(1):123-130

[8] Kappus N, Weinstock-Guttman B, Hagemeier J, Kennedy C, Melia R, Carl E, et al. Cardiovascular risk factors are associated with increased lesion burden and brain atrophy in multiple sclerosis. Journal of Neurology, Neurosurgery, and Psychiatry. 2015; **87**(2):181-187

[9] Marrie RA, Horwitz R, Cutter G, Tyry T, Campagnolo D, Vollmer T. Comorbidity delays diagnosis and increases disability at diagnosis in MS. Neurology. 2009;**72**(2):117-124

[10] Moccia M, Lanzillo R, Palladino R, Maniscalco GT, De Rosa A, Russo C, et al. The Framingham cardiovascular risk score in multiple sclerosis. European Journal of Neurology. 2015; **22**(8):1176-1183

[11] Farez MF, Fiol MP, Gaitán MI, Quintana FJ, Correale J. Sodium intake is associated with increased disease activity in multiple sclerosis. Journal of Neurology, Neurosurgery, and Psychiatry. 2015;**86**(1):26-31

[12] Ranadive SM, Yan H, Weikert M, Lane AD, Linden MA, Baynard T, et al. Vascular dysfunction and physical activity in multiple sclerosis. Medice and Science Sport and Exercise. 2012;**44**(2):238-243

[13] Yuksel B, Koc P, Ozaydin Goksu E, Karacay E, Kurtulus F, Cekin Y, et al. Is multiple sclerosis a risk factor for atherosclerosis? Journal of Neuroradiology. 2021;**48**(2):99-103

[14] Pelizzari L, Jakimovski D, Laganà MM, Bergsland N, Hagemeier J, Baselli G, et al. Five-year longitudinal study of neck vessel cross-sectional area in multiple sclerosis. American Journal of Neuroradiology. 2018;**39**(9): 1703-1709

[15] Zamboni P, Galeotti R, Menegatti E, Malagoni AM, Tacconi G, Dall'Ara S, et al. Chronic cerebrospinal venous

insufficiency in patients with multiple sclerosis. Journal of Neurology, Neurosurgery, and Psychiatry. 2009;**80**(4):392-399

[16] Sethi SK, Daugherty AM, Gadda G, Utriainen DT, Jiang J, Raz N, et al. Jugular anomalies in multiple sclerosis are associated with increased collateral venous flow. American Journal of Neuroradiology. 2017;**38**(8):1617-1622

[17] Singh AV, Zamboni P. Anomalous venous blood flow and iron deposition in multiple sclerosis. Journal of Cerebral Blood Flow and Metabolism. 2009; **29**(12):1867-1878

[18] Zamboni P, Menegatti E, Galeotti R, Malagoni AM, Tacconi G, Dall'Ara S, et al. The value of cerebral Doppler venous haemodynamics in the assessment of multiple sclerosis. Journal of the Neurological Sciences. 2009; **282**(1-2):21-27

[19] Beggs CB. Venous hemodynamics in neurological disorders: An analytical review with hydrodynamic analysis. BMC Medicine. 2013;**11**(1):142

[20] Haacke EM, Feng W, Utriainen D, Trifan G, Wu Z, Latif Z, et al. Patients with multiple sclerosis with structural venous abnormalities on MR imaging exhibit an abnormal flow distribution of the internal jugular veins. Journal of Vascular and Interventional Radiology. 2012;**23**(1):60-63

[21] Krsmanović Ž, Živković M, Lepić T, Stanković A, Raičević R, Dinčić E. Small internal jugular veins with restricted outflow are associated with severe multiple sclerosis: a sonographer-blinded, case–control ultrasound study. BMC Neurology. 2013;**13**(1):90

[22] Cocozza S, Canna A, Lanzillo R, Russo C, Postiglione E, Liuzzi R, et al. Lack of correlation between extracranial venous abnormalities and multiple sclerosis: A quantitative MRI study. The British Journal of Radiology. 2016; **89**(1064):20160321

[23] Zivadinov R, Lopez-Soriano A, Weinstock-Guttman B, Schirda CV, Magnano CR, Dolic K, et al. Use of MR venography for characterization of the extracranial venous system in patients with multiple sclerosis and healthy control subjects. Radiology. 2011;**258**(2):562-570

[24] Driedger SM, Maier R, Marrie RA, Brouwers M. Caught in a no-win situation: Discussions about CCSVI between persons with multiple sclerosis and their neurologists – a qualitative study. BMC Neurology. 2017;**17**(1):176

[25] Traboulsee AL, Machan L, Girard JM, Raymond J, Vosoughi R, Hardy BW, et al. Safety and efficacy of venoplasty in MS: A randomized, double-blind, sham-controlled phase II trial. Neurology. 2018;**91**(18): e1660-e1668

[26] Tsivgoulis G, Faissner S, Voumvourakis K, Katsanos AH, Triantafyllou N, Grigoriadis N, et al. "Liberation treatment" for chronic cerebrospinal venous insufficiency in multiple sclerosis: The truth will set you free. Brain and Behavior: A Cognitive Neuroscience Perspective. 2015; **5**(1):3-12

[27] Jagannath VA, Pucci E, Asokan GV, Robak EW. Percutaneous transluminal angioplasty for treatment of chronic cerebrospinal venous insufficiency (CCSVI) in people with multiple sclerosis. Cochrane Database of Systematic Reviews. 2019;**5**(5): CD009903

[28] Zamboni P, Galeotti R, Salvi F, Giaquinta A, Setacci C, Alborino S, et al. Effects of venous angioplasty on cerebral lesions in multiple sclerosis: Expanded analysis of the brave dreams double-blind, sham-controlled randomized trial. Journal of

Endovascular Therapy. 2020;**27**(1): 1526602819890110

[29] Filippi M, Rocca MA, Barkhof F, Brück W, Chen JT, Comi G, et al. Association between pathological and MRI findings in multiple sclerosis. Lancet Neurology. 2012;**11**(4):349-360

[30] Öner S, Kahraman AS, Özcan C, Özdemir ZM, Ünlü S, Kamışlı Ö, et al. Cerebrospinal fluid dynamics in patients with multiple sclerosis: The role of phase-contrast MRI in the differential diagnosis of active and chronic disease. Korean Journal of Radiology. 2018; **19**(1):72-78

[31] Swank RL, Roth JG, Woody DCJ. Cerebral blood flow and red cell delivery in normal subjects and in multiple sclerosis. Neurological Research. 1983;**5**(1):37-59

[32] Brooks DJ, Leenders KL, Head G, Marshall J, Legg NJ, Jones T. Studies on regional cerebral oxygen utilisation and cognitive function in multiple sclerosis. Journal of Neurology, Neurosurgery, and Psychiatry. 1984;**47**(11):1182-1191

[33] Lycke J, Wikkelsö C, Bergh AC, Jacobsson L, Andersen O. Regional cerebral blood flow in multiple sclerosis measured by single photon emission tomography with technetium-99m hexamethylpropyleneamine oxime. European Neurology. 1993;**33**(2): 163-167

[34] Sun X, Tanaka M, Kondo S, Okamoto K, Hirai S. Clinical significance of reduced cerebral metabolism in multiple sclerosis: A combined PET and MRI study. Annals of Nuclear Medicine. 1998;**12**(2):89-94

[35] Novak V. Cognition and hemodynamics. Current Cardiovascular Risk Reports. 2012;**6**(5):380-396

[36] Turner MP, Hubbard NA, Sivakolundu DK, Himes LM, Hutchison JL, Hart J, et al. Preserved canonicality of the BOLD hemodynamic response reflects healthy cognition: Insights into the healthy brain through the window of multiple sclerosis. NeuroImage. 2019;**190**:46-55

[37] Marshall O, Chawla S, Lu H, Pape L, Ge Y. Cerebral blood flow modulation insufficiency in brain networks in multiple sclerosis: A hypercapnia MRI study. Journal of Cerebral Blood Flow and Metabolism. 2016;**36**(12):2087-2095

[38] Abbott NJ, Rönnbäck L, Hansson E. Astrocyte-endothelial interactions at the blood-brain barrier. Nature Reviews Neuroscience. 2006;**7**(1):41-53

[39] Lassmann H, van Horssen J, Mahad D. Progressive multiple sclerosis: Pathology and pathogenesis. Nature Reviews Neurology. 2012;**8**(11):647-656

[40] Miller DH, Barkhof F, Frank JA, Parker GJM, Thompson AJ. Measurement of atrophy in multiple sclerosis: Pathological basis, methodological aspects and clinical relevance. Brain. 2002;**125**(Pt 8): 1676-1695

[41] Law M, Saindane AM, Ge Y, Babb JS, Johnson G, Mannon LJ, et al. Microvascular abnormality in relapsing-remitting multiple sclerosis: Perfusion MR imaging findings in normal-appearing white matter. Radiology. 2004;**231**(3):645-652

[42] Wuerfel J, Bellmann-Strobl J, Brunecker P, Aktas O, McFarland H, Villringer A, et al. Changes in cerebral perfusion precede plaque formation in multiple sclerosis: A longitudinal perfusion MRI study. Brain. 2004; **127**(Pt 1):111-119

[43] Ge Y, Law M, Johnson G, Herbert J, Babb JS, Mannon LJ, et al. Dynamic susceptibility contrast perfusion MR imaging of multiple sclerosis lesions: Characterizing hemodynamic

impairment and inflammatory activity. AJNR. American Journal of Neuroradiology. 2005;**26**(6):1539-1547

[44] Sowa P, Bjørnerud A, Nygaard GO, Damangir S, Spulber G, Celius EG, et al. Reduced perfusion in white matter lesions in multiple sclerosis. European Journal of Radiology. 2015;**84**(12): 2605-2612

[45] Peruzzo D, Castellaro M, Calabrese M, Veronese E, Rinaldi F, Bernardi V, et al. Heterogeneity of cortical lesions in multiple sclerosis: An MRI perfusion study. Journal of Cerebral Blood Flow and Metabolism. 2013;**33**(3):457-463

[46] Narayana PA, Zhou Y, Hasan KM, Datta S, Sun X, Wolinsky JS. Hypoperfusion and T1-hypointense lesions in white matter in multiple sclerosis. Multiple Sclerosis. 2014;**20**(3):365-373

[47] Inglese M, Park S-J, Johnson G, Babb JS, Miles L, Jaggi H, et al. Deep gray matter perfusion in multiple sclerosis: Dynamic susceptibility contrast perfusion magnetic resonance imaging at 3 T. Archives of Neurology. 2007;**64**(2):196-202

[48] Rashid W, Parkes LM, Ingle GT, Chard DT, Toosy AT, Altmann DR, et al. Abnormalities of cerebral perfusion in multiple sclerosis. Journal of Neurology, Neurosurgery, and Psychiatry. 2004;**75**(9):1288-1293

[49] Adhya S, Johnson G, Herbert J, Jaggi H, Babb JS, Grossman RI, et al. Pattern of hemodynamic impairment in multiple sclerosis: Dynamic susceptibility contrast perfusion MR imaging at 3.0 T. NeuroImage. 2006;**33**(4):1029-1035

[50] Varga AW, Johnson G, Babb JS, Herbert J, Grossman RI, Inglese M. White matter hemodynamic abnormalities precede sub-cortical gray matter changes in multiple sclerosis. Journal of the Neurological Sciences. 2009;**282**(1-2):28-33

[51] Saindane AM, Law M, Ge Y, Johnson G, Babb JS, Grossman RI. Correlation of diffusion tensor and dynamic perfusion MR imaging metrics in normal-appearing corpus callosum: Support for primary hypoperfusion in multiple sclerosis. AJNR. American Journal of Neuroradiology. 2007; **28**(4):767-772

[52] Hojjat S-P, Kincal M, Vitorino R, Cantrell CG, Feinstein A, Zhang L, et al. Cortical perfusion alteration in normal-appearing gray matter is most sensitive to disease progression in relapsing-remitting multiple sclerosis. AJNR. American Journal of Neuroradiology. 2016;**37**(8):1454-1461

[53] Hostenbach S, Raeymaekers H, Van Schuerbeek P, Vanbinst A-M, Cools W, De Keyser J, et al. The role of cerebral hypoperfusion in multiple sclerosis (ROCHIMS) trial in multiple sclerosis: Insights from negative results. Frontiers in Neurology. 2020;**11**:674

[54] Debernard L, Melzer TR, Van Stockum S, Graham C, Wheeler-Kingshott CA, Dalrymple-Alford JC, et al. Reduced grey matter perfusion without volume loss in early relapsing-remitting multiple sclerosis. Journal of Neurology, Neurosurgery, and Psychiatry. 2014;**85**(5):544-551

[55] Doche E, Lecocq A, Maarouf A, Duhamel G, Soulier E, Confort-Gouny S, et al. Hypoperfusion of the thalamus is associated with disability in relapsing remitting multiple sclerosis. Journal of Neuroradiology. 2017;**44**(2):158-164

[56] Bester M, Forkert ND, Stellmann JP, Stürner K, Aly L, Drabik A, et al. Increased perfusion in normal appearing white matter in high inflammatory multiple sclerosis

patients. PLoS One. 2015;**10**(3): e0119356

[57] Zhang X, Guo X, Zhang N, Cai H, Sun J, Wang Q, et al. Cerebral blood flow changes in multiple sclerosis and neuromyelitis optica and their correlations with clinical disability. Frontiers in Neurology. 2018;**2**:9

[58] Efendi H. Clinically isolated syndromes: Clinical characteristics, differential diagnosis, and management. Noro Psikiyatri Arsivi. 2015;**52**(Suppl. 1): S1-S11

[59] Miller D, Barkhof F, Montalban X, Thompson A, Filippi M. Clinically isolated syndromes suggestive of multiple sclerosis, part 2: Non-conventional MRI, recovery processes, and management. Lancet Neurology. 2005;**4**(6):341-348

[60] Haselhorst R, Kappos L, Bilecen D, Scheffler K, Möri D, Radü EW, et al. Dynamic susceptibility contrast MR imaging of plaque development in multiple sclerosis: Application of an extended blood-brain barrier leakage correction. Journal of Magnetic Resonance Imaging. 2000;**11**(5):495-505

[61] Papadaki EZ, Mastorodemos VC, Amanakis EZ, Tsekouras KC, Papadakis AE, Tsavalas ND, et al. White matter and deep gray matter hemodynamic changes in multiple sclerosis patients with clinically isolated syndrome. Magnetic Resonance in Medicine. 2012;**68**(6):1932-1942

[62] Paling D, Thade Petersen E, Tozer DJ, Altmann DR, Wheeler-Kingshott CAM, Kapoor R, et al. Cerebral arterial bolus arrival time is prolonged in multiple sclerosis and associated with disability. Journal of Cerebral Blood Flow and Metabolism. 2014;**34**(1):34-42

[63] D'haeseleer M, Hostenbach S, Peeters I, Sankari S El, Nagels G, De Keyser J, et al. Cerebral hypoperfusion: A new pathophysiologic concept in multiple sclerosis? Journal of Cerebral Blood Flow and Metabolism 2015;**35**(9):1406-1410.

[64] Mezei Z, Olah L, Kardos L, Kovacs RK, Csiba L, Csepany T. Cerebrovascular hemodynamic changes in multiple sclerosis patients during head-up tilt table test: Effect of high-dose intravenous steroid treatment. Journal of Neurology. 2013;**260**(9):2335-2342

[65] Adamec I, Crnošija L, Junaković A, Krbot Skorić M, Habek M. Progressive multiple sclerosis patients have a higher burden of autonomic dysfunction compared to relapsing remitting phenotype. Clinical Neurophysiology. 2018;**129**(8):1588-1594

[66] Racosta JM. Autonomic nervous system dysfunction and fatigue in multiple sclerosis: Common pathophysiology or spurious association? Clinical Autonomic Research. 2019;**29**(3):261-262

[67] Valencia-Sanchez C, Goodman BP, Carter JL, Wingerchuk DM. The spectrum of acute cardiopulmonary events associated with multiple sclerosis exacerbations. Multiple Sclerosis. 2019;**25**(6):758-765

[68] Findling O, Hauer L, Pezawas T, Rommer PS, Struhal W, Sellner J. Cardiac autonomic dysfunction in multiple sclerosis: A systematic review of current knowledge and impact of immunotherapies. Journal of Clinical Medicine. 2020;**9**(2):335

[69] De Keyser J, Steen C, Mostert JP, Koch MW. Hypoperfusion of the cerebral white matter in multiple sclerosis: Possible mechanisms and pathophysiological significance. Journal of Cerebral Blood Flow and Metabolism. 2008;**28**(10):1645-1651

[70] Chiaravalloti ND, DeLuca J. Cognitive impairment in multiple

sclerosis. Lancet Neurology. 2008; **7**(12):1139-1151

[71] Zamboni P. Chapter 29 - The Contribution of Extra Cranial Venous Drainage to Neuro-Inflammation in Multiple Sclerosis. In: Minagar A, editor. Neuroinflammation. Second ed. Massachusetts: Academic Press; 2018. pp. 579-599

[72] Kutzelnigg A, Lucchinetti CF, Stadelmann C, Brück W, Rauschka H, Bergmann M, et al. Cortical demyelination and diffuse white matter injury in multiple sclerosis. Brain. 2005;**128**(Pt 11):2705-2712

[73] Moffett JR, Ross B, Arun P, Madhavarao CN, Namboodiri AMA. N-Acetylaspartate in the CNS: From neurodiagnostics to neurobiology. Progress in Neurobiology. 2007; **81**(2):89-131

[74] Steen C, D'haeseleer M, Hoogduin JM, Fierens Y, Cambron M, Mostert JP, et al. Cerebral white matter blood flow and energy metabolism in multiple sclerosis. Multiple Sclerosis. 2013;**19**(10):1282-1289

[75] Vucic S, Burke T, Lenton K, Ramanathan S, Gomes L, Yannikas C, et al. Cortical dysfunction underlies disability in multiple sclerosis. Multiple Sclerosis. 2012;**18**(4):425-432

[76] D'haeseleer M, Cambron M, Vanopdenbosch L, De Keyser J. Vascular aspects of multiple sclerosis. Lancet Neurology. 2011;**10**(7):657-666

[77] Haider L, Zrzavy T, Hametner S, Höftberger R, Bagnato F, Grabner G, et al. The topograpy of demyelination and neurodegeneration in the multiple sclerosis brain. Brain. 2016;**139** (Pt 3):807-815

[78] Stickland R, Allen M, Magazzini L, Singh KD, Wise RG, Tomassini V. Neurovascular coupling during visual stimulation in multiple sclerosis: A MEG-fMRI Study. Neuroscience. 2019;**403**:54-69

[79] Lu F, Selak M, O'Connor J, Croul S, Lorenzana C, Butunoi C, et al. Oxidative damage to mitochondrial DNA and activity of mitochondrial enzymes in chronic active lesions of multiple sclerosis. Journal of the Neurological Sciences. 2000;**177**(2):95-103

[80] Lassmann H. Hypoxia-like tissue injury as a component of multiple sclerosis lesions. Journal of the Neurological Sciences. 2003;**206**(2): 187-191

[81] Barnett MH, Prineas JW. Relapsing and remitting multiple sclerosis: Pathology of the newly forming lesion. Annals of Neurology. 2004;**55**(4): 458-468

[82] Einarsdottir AB, Olafsdottir OB, Hjaltason H, Hardarson SH. Retinal oximetry is affected in multiple sclerosis. Acta Ophthalmologica. 2018;**96**(5):528-530

[83] Mulligan SJ, BA MV. Calcium transients in astrocyte endfeet cause cerebrovascular constrictions. Nature. 2004;**431**(7005):195-199

[84] Choi I-Y, Lee S-P, Denney DR, Lynch SG. Lower levels of glutathione in the brains of secondary progressive multiple sclerosis patients measured by 1H magnetic resonance chemical shift imaging at 3 T. Multiple Sclerosis. 2011;**17**(3):289-296

[85] Tasset I, Agüera E, Sánchez-López F, Feijóo M, Giraldo AI, Cruz AH, et al. Peripheral oxidative stress in relapsing-remitting multiple sclerosis. Clinical Biochemistry. 2012; **45**(6):440-444

[86] Majeed AQ, Hasan ZN, Jalal MF. Magnetic resonance venography findings in a group of patients with

multiple sclerosis. Multiple Sclerosis and Related Disorders. 2018;**26**:248

[87] Jahng G-H, Li K-L, Ostergaard L, Calamante F. Perfusion magnetic resonance imaging: A comprehensive update on principles and techniques. Korean Journal of Radiology. 2014; **15**(5):554-577

[88] Schega L, Hamacher D, Sailer M, Broscheid K-C. Reliability of the hemodynamic response during walking in people with multiple sclerosis: An fNIRS study. Archives of Physical Medicine and Rehabilitation. 2019; **100**(10):e115

[89] Lapointe E, Li DKB, Traboulsee AL, Rauscher A. What have we learned from perfusion MRI in multiple sclerosis? AJNR. American Journal of Neuroradiology. 2018;**39**(6):994-1000

[90] Sheng H, Zhao B, Ge Y. Blood perfusion and cellular microstructural changes associated with iron deposition in multiple sclerosis lesions. Frontiers in Neurology. 2019;**10**:747

[91] Gaillard F, Lee M. Dynamic susceptibility contrast (DSC) MR perfusion. Radiopaedia.org. Radiopaedia.org; 2016 [cited 2021 Oct 8]. Available from: http://radiopaedia.org/articles/43780

[92] Sisco NJ, Borazanci A, Dortch R, Stokes AM. Investigating the relationship between multi-scale perfusion and white matter microstructural integrity in patients with relapsing remitting MS. Multiple Sclerosis Journal Experimental Translational and Clinical. 2021;7(3): 20552173211037002

[93] Cho J, Nguyen TD, Huang W, Sweeney EM, Luo X, Kovanlikaya I, et al. Brain oxygen extraction fraction mapping in patients with multiple sclerosis. Journal of Cerebral Blood Flow and Metabolism. Sep 2021: 271678X211048031

[94] West K, Sivakolundu D, Maruthy G, Zuppichini M, Liu P, Thomas B, et al. Baseline cerebral metabolism predicts fatigue and cognition in multiple sclerosis patients. NeuroImage Clinical. 2020;**27**:102281

[95] Lattanzi S, Carbonari L, Pagliariccio G, Cagnetti C, Luzzi S, Bartolini M, et al. Predictors of cognitive functioning after carotid revascularization. Journal of the Neurological Sciences. 2019;**405**:116435

[96] Wahls T, Scott MO, Alshare Z, Rubenstein L, Darling W, Carr L, et al. Dietary approaches to treat MS-related fatigue: comparing the modified paleolithic (wahls elimination) and low saturated fat (Swank) diets on perceived fatigue in persons with relapsing-remitting multiple sclerosis: Study protocol for a randomized control. Trials. 2018;**19**(1):309

[97] Shahrampour S, Heholt J, Wang A, Vedaei F, Mohamed FB, Alizadeh M, et al. N-acetyl cysteine administration affects cerebral blood flow as measured by arterial spin labeling MRI in patients with multiple sclerosis. Heliyon. 2021; 7(7):e07615

Chapter 6

The Shear Stress/KLF2/Nrf2/ARE Pathway: A Hemodynamic Defense against Oxidative Stress

John M. Owen and Kenneth J. Dormer

Abstract

Many diseases have oxidative stress and inflammation as underlying pathological features, including metabolic and inflammatory/autoimmune disorders, diseases of the lung, liver, kidney, gastrointestinal tract, cardiovascular and nervous systems. A leading physiological mechanism for oxidative stress is the nuclear erythroid-related factor 2-like 2/antioxidant response element (Nrf2/ARE) signaling pathway. It maintains intracellular homeostasis and protects cells from oxidative damage by inducing phase II detoxifying and oxidative-stress responsive genes. Nrf2 transcription factor functions as the key controller of the redox homeostatic gene regulatory network, and is tightly controlled by the repressor protein, Kelch-like ECH-associated protein 1 (Keap1). Pharmacological agents to inhibit Keap1 and boost effectiveness of the Nrf2/ARE pathway have been developed and more are in development. This chapter elucidates the importance of hemodynamic laminar shear stress in oxidative homeostasis and examines hemodynamic induction of the shear stress (SS)/Krupple-like factor2 (KLF2) /Nrf2/ARE pathway as a means to combat oxidative stress through hemodynamics.

Keywords: shear stress, mechanotransduction, Nrf2, KLF2, oxidative stress, hemodynamics, homeostasis

1. Introduction

In the mid-nineteenth century, Claude Bernard introduced the idea of the "inner world" when he theorized that bodily systems function to maintain a constant internal environment—what he called the *milieu intérieur*. A half century later, Walter Cannon popularized the concept in his book, Spirit of the Body, wherein he coined the word *homeostasis,* which, "...does not imply something set and immobile, a stagnation. It means a condition—a condition which may vary, but is relatively constant." Homeostasis is achieved by constant rebalancing within the body of competing mechanisms such as vasodilation vs. vasoconstriction, coagulation vs. anti-coagulation, and inflammatory vs. anti-inflammatory elements [1]. This balance of control is usually attained by negative feedback mechanisms [2].

All imbalances of these antagonistic mechanisms lead to heterostasis or disease, which are deleterious to the body in some respect. None perhaps is more harmful to the body and more likely to lead to morbidity and mortality than excess oxidative stress with inadequate antioxidant response. Oxidative stress-based diseases affect all parts of the body and manifest themselves through some of the worst diseases

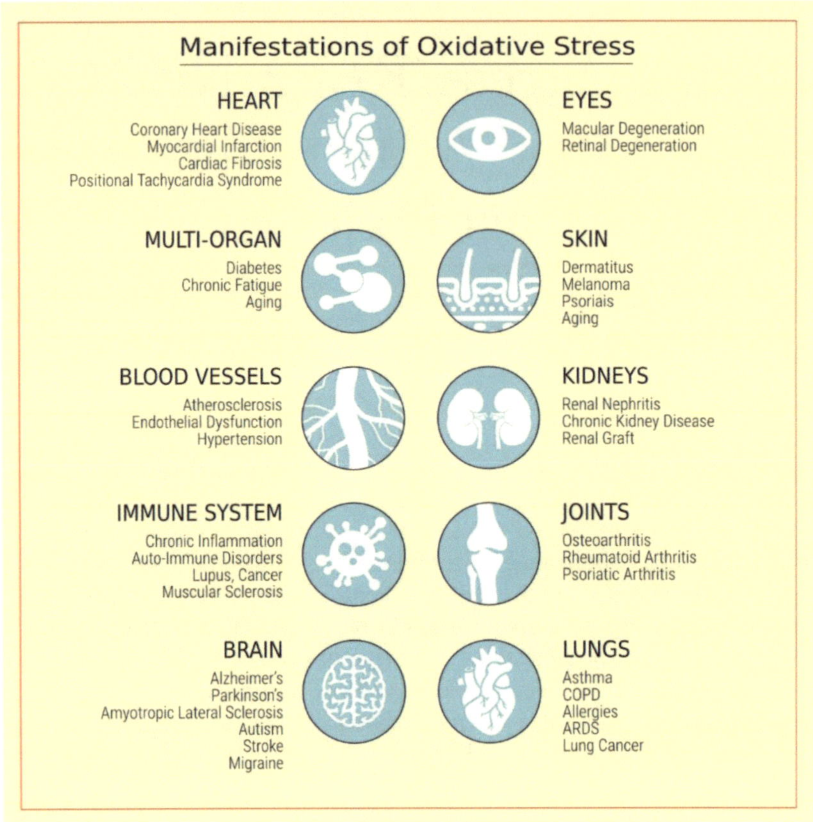

Figure 1.
Manifestations of oxidative stress. *Oxidative stress has been identified as a causative or contributing factor in many diseases affecting most of the body's major organs. Thirty-five examples of oxidative diseases are shown for eight organ systems.*

which afflict mankind. To compound the problem, these diseases currently have only symptomatic relief (**Figure 1**).

Hemodynamics play a key role in the SS/KLF2/Nrf2/ARE antioxidant pathway since its resulting shear stress upregulates both KLF2 and Nrf2 along with other antioxidants such as catalase, superoxide dismutase, glutathione peroxidase, sirtuin, bilirubin, and others [3, 4]. Each of these substances add to the collective antioxidant production generated through this pathway. The SS/KLF2/Nrf2/ARE pathway is only a part of a broad biochemical cascade induced by shear stress that creates a complex, multifaceted, overlapping and interacting response to oxidative stress which contributes to oxidation/reduction (Redox) homeostasis.

2. The endothelium

Endothelial cells (EC) comprise the endothelium, lining the lumen of all blood vessels. The endothelium is the largest organ in the human body with a total weight comparable to other vital organs and possessing a surface area larger than six tennis courts [5]. Once thought of as a passive barrier, now viewed as an organ crucial to maintaining vascular health, endothelial dysfunction is an important factor in the initiation, progression and clinical complication of vascular disease. EC are an integral part of tissues and organs. A unique cellular system lining the inside of blood vessels, the EC form an interface between circulating blood and the parenchymal cells. EC are regulators of hemostasis, vasomotor control, immunological and platelet functions, inflammatory responses, vascular smooth muscle cell growth and migration. EC fundamentally control vascular tone by sensing and reacting through secretion of transcellular and intracellular signaling molecules. Additionally, the endothelium forms an essential vascular barrier for solute transport and osmotic balance [6, 7].

The endothelium is easily overlooked in clinical practice since it does not lend itself to evaluation. Compounding the problem is that many physicians are not trained in endothelial health. Few textbooks focus on EC and medical school curricula generally lack courses on the endothelium. Additionally, while diseases of other organs are associated with measurable biomarkers, endothelial dysfunction has no reliable markers. Like other organs in the body, the endothelium is highly complex with physiological, biochemical and biomechanical parameters. The endothelium, more than most tissues in the body, is adaptive and flexible, responding to the ever-changing milieu of the local microenvironment.

Unsurprisingly, the EC have a broad potential as therapeutic targets. Since EC are strategically located between the blood and tissue, they are rapidly exposed to biomolecules, injected pharmacological agents, as well as hemodynamic physical forces. Also, the endothelium is highly changeable in size and elasticity in response to intrinsic or extrinsic physiological controllers, and thus is amenable to therapeutic intervention while supplying a direct line of communication with every organ in the body [8].

3. Hemodynamics, shear stress and mechanotransduction

Mechanical forces guide the form and function of the cardiovascular system, whose main role of transporting blood to every tissue in the body is essentially physico-mechanical. The stroke volume generates hemodynamic forces on the arterial vasculature: wall shear stress, hydrostatic pressures and cyclic stretch [9]. Laminar SS, the more important of these forces, is a tangential force arising due to the dragging friction of blood elements with the vessel wall. SS will vary from a low of ~1 dyne/cm^2 in veins up to >50 dyne/cm^2 in arterial vessels [10]. Hemodynamically driven blood flow and SS produce EC mechanotransduction, a group of events whereby a cell can actively sense, integrate, and convert a physical stimulus into electrical and biochemical signals [11]. These forces are sensed and interpreted by EC in the luminal vessel wall to: a) guide development during embryogenesis and remodeling during postnatal and adult life; b) optimize blood flow to the tissues and; c) ensure mechanical integrity of the vessel walls. These SS signals bring about intracellular changes, such as activation of signaling pathways and transcriptional regulation that modify gene and protein expressions as well as endothelial phenotype and function [12]. Shear stress-induced mechanotransduction influences key molecules and signaling pathways that lead to the changes in cell functions and behavior.

EC are exposed to fluid forces of greater magnitude than those experienced by other tissues. The mechanically related responses controlled by the endothelium are most important in the control of vascular tone in regulating blood flow. The principal functions of endothelium include: a) maintenance of anticoagulant properties; b) regulation of vascular permeability; c) control of vessel diameter; and d) responses to pathological consequences associated with inflammation, wound healing, and cardiovascular disorders.

Hemodynamic factors in these processes can influence endothelial anatomy and function either by the direct action of shear stress and other stretch forces on the endothelium or by indirect modification of the local concentrations of chemicals and agonists at the endothelial surface (**Figure 2**). The mechanisms may have overlapping actions such as direct forces acting on surface enzymes while receptors concurrently modify enzyme-substrate and agonist-receptor interactions while one or both can be influenced by convective or diffusive transport [13].

It has been shown that EC-induced gene expression is important in hemostasis, thrombosis, growth regulation and proinflammatory activation and is transcriptionally regulated by mechanotransduction [4]. Many of these activated regulatory genes are directly involved in EC adhesion (e.g., ICAM-1). These observations suggest a novel paradigm linking biomechanical stimulation with endothelial activation. Studies have revealed the existence of shear stress response elements (SSRE) in the promoters of physiologically relevant genes such as the platelet-derived growth factor (PDGF), endothelial nitric oxide synthase (eNOS) and vascular cellular adhesion molecule (VCAM-1), that act to up- or down-regulate gene transcription.

One of the key shear stress-generated endothelial molecules is eNOS. This enzyme generates nitric oxide (NO) from L-arginine and O2. NO regulates EC survival, vascular tone (vasodilation), angiogenesis and possesses anti-inflammatory and antioxidant properties [14].

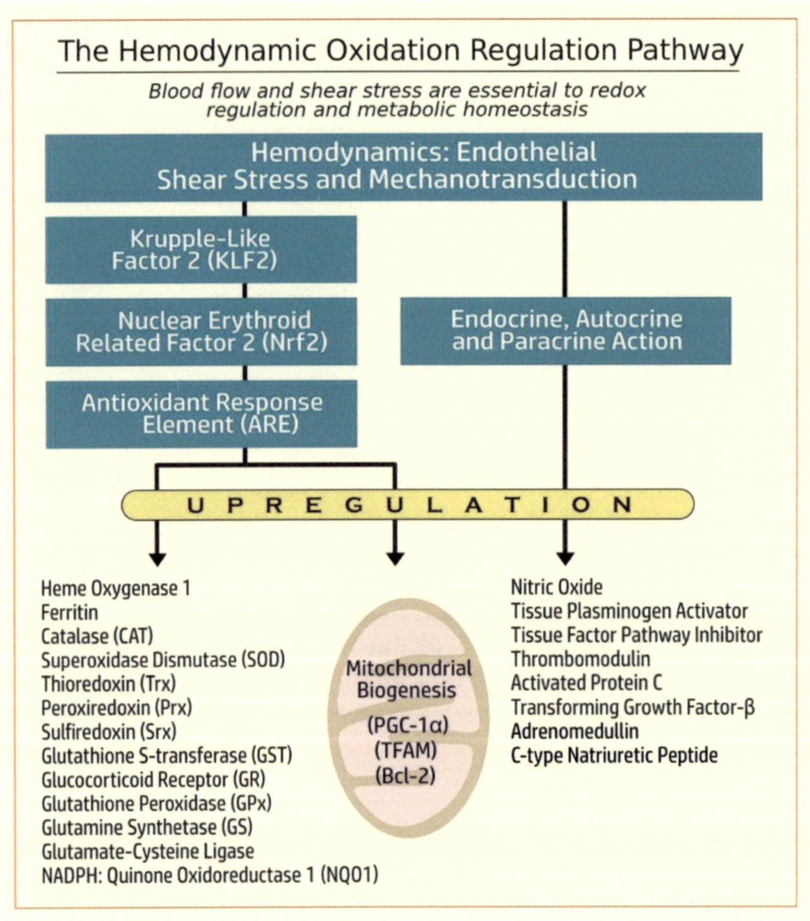

Figure 2.
The hemodynamic oxidation regulation pathway*. Hemodynamic shear stress and mechanotransduction create a cascade of complex, interacting events which induce dozens of physiologically active substances. Some are brought about through the SS/KLF2/Nrf2/ARE pathway, others are through direct autocrine and paracrine pathways.*

To date, few have attempted to use hemodynamics, to counter excess oxidative stress. This is in part due to a shortage of methods with which to engage the body hemodynamically. Existing therapeutic methods of increasing blood flow sufficiently to generate therapeutic levels of shear stress include: exercise, electrical stimulation, external counter pulsation (ECP), periodic acceleration, and simulated (passive) skeletal muscle exercise.

Exercise is an excellent therapy to combat oxidative stress even though it generates reactive oxygen species (ROS). Unfortunately, most patients in need of this therapy either cannot, or will not, comply with prescriptive exercise needed to reach therapeutic levels of SS. Electrical stimulation has been used for years to simulate exercise with good success, but the field suffers from confusing heterogeneity. There is an overwhelming number of electrical stimulators in the market with differences in voltage, current, waveform, protocol, size and number of electrodes employed. Also, there are not many randomized clinical trials of electrical stimulation with statistically validated results. Other alternatives to exercise include a simulated jogging device that provides passive cycles of leg movement resulting in an increased level of blood flow and shear stress in the legs. There are also external counter pulsation (ECP) devices that produce SS. Their use, however, is currently limited to treating angina pectoris and heart failure. ECP consists of pneumatic cuffs placed on legs and lower torso with cyclic inflations and deflations that are timed to the patient's heartbeat such that cuffs inflate at the beginning of diastole and deflate at the beginning of systole. In theory, this action increases the number of pulses in the circulation thereby producing additional SS. Lastly, SS has been increased through a motorized bed that has a reciprocating motion, creating back-and-forth movement of blood flow. This therapy, known as whole-body periodic acceleration, does not increase blood circulation, however it does increase levels of shear stress [15]. The use of hemodynamics as a therapy, despite today's lack of strong evidence, appears to hold promise, with Fledderus and colleagues [16] finding that, "Physiological levels of shear stress will induce activation and nuclear translocation of Nrf2, and Nrf2-dependent cytoprotective gene expression." They also found that "SS generated KLF2 primes the activation of the Nrf2 pathway by inducing nuclear localization of Nrf2".

4. Oxidative stress and homeostasis

Excessive, chronic oxidative stress has been implicated in the development and exacerbation of diseases affecting most of the body's major organs, including cancer, diabetes, autoimmune, cutaneous, neurodegenerative, pulmonary and cardiovascular diseases, infection, inflammation, and aging (**Figure 1**). Endogenous oxidative stressors normally result from metabolic processes involving mitochondria. Chronic exposure to excess reactive oxygen species causes cellular and macromolecular damage [17]. Oxidative stress is the result of an imbalance of pro-oxidant and antioxidant substances that lead to the generation of toxic ROS, such as hydrogen peroxide, nitric oxide, superoxide, hydroxyl radicals and others [18]. The production of ROS is usually in balance with homeostatic antioxidant enzymes such as superoxide dismutase (SOD), catalase (CAT) and glutathione peroxidase (Gpx). *In vivo* studies have found that most oxidative damage occurs from reduced levels of antioxidants rather than increased ROS production [19]. Adequate levels of both are considered to be essential for normal cell function. Mitochondria create their own antioxidants to balance their generation of ROS, such as manganese SOD (Mn-SOD) which converts O2•- to H2O2 which is further reduced by CAT and Gpx to harmless H2O and O2. Importantly, CAT, Gpx, and SOD are essential antioxidants that are hemodynamically upregulated in response to physiological levels of endothelial SS [20, 21]. The Copper/zinc-SOD (Cu/Zn-SOD) antioxidant, which is also upregulated in response to shear stress, plays a role in stabilizing O2•- and contributes to the homeostatic redox state. Antioxidant defenses are extremely important as they eliminate free radicals, thereby providing biological protection. These systems not only defend against the problems of oxidative damage but are essential for disease prevention [22].

Oxidant and antioxidant signaling are both features of oxidative homeostasis, which is the maintenance of nucleophilic tone and a healthy physiological steady state. Redox imbalance is rapidly reversed by feedback reactions, maintained by continuous signaling for production and elimination of electrophiles and nucleophiles, thus maintaining homeostasis [23]. The production of oxygen free radicals sometimes exceeds the capacity of the endogenous antioxidant system and oxidative stress occurs as well as cellular injury. Oxygen free radicals can cause cellular membrane lipid peroxidation and protein oxidation which leads to disruption of cellular integrity. In addition, apoptosis and autophagy, resulting from oxidative stress, represent important mechanisms that can lead to the destruction of cells in many systems [24].

Hemodynamic SS is a key modulator of the body's response to oxidative stress. Physiological levels of laminar SS, ~12 to 15 dynes/cm^2, as in laminar arterial flow, promote EC survival and quiescence, alignment of EC in the direction of flow, and secretion of substances that reduce oxidation and coagulation while allowing

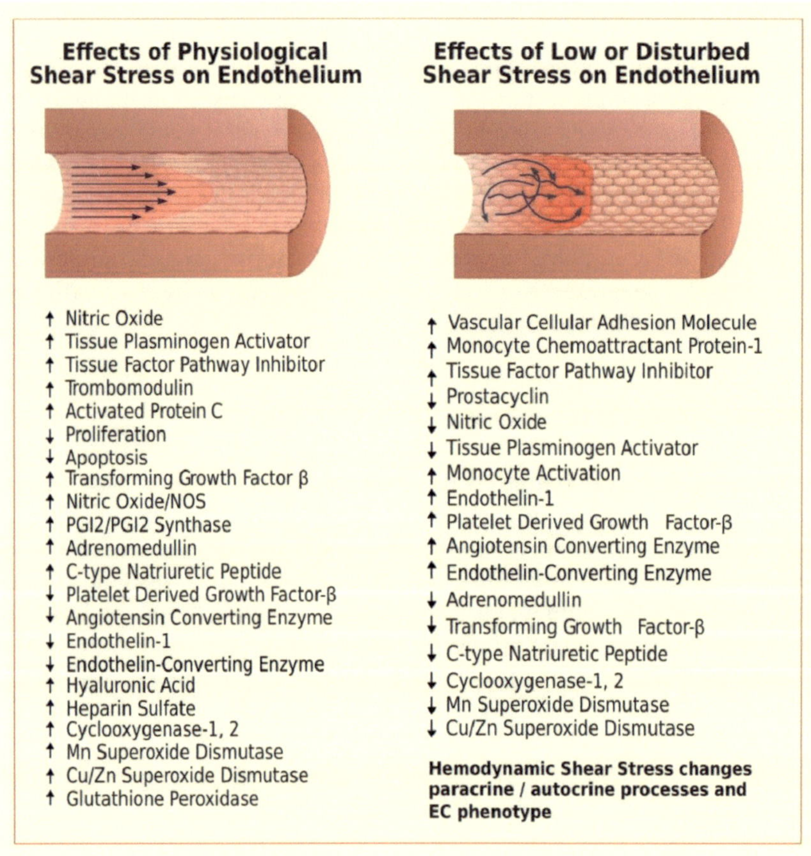

Figure 3.
Effects of shear stress on endothelium. *Biochemical changes brought about through hemodynamics, shear stress and mechanotransduction can have both beneficial and destructive consequences throughout the body. Generally, low or disturbed flow and SS lead to endothelial dysfunction and disease, while normal physiological flow helps maintain endothelial homeostasis.*

vasodilation to increase flow. In contrast, low SS, or turbulent flow and other atypical flow patterns that involve changes in direction and magnitude of flow with shear stress <5 dynes/cm^2, promote ROS generation, adhesive and inflammatory molecules, vasoconstrictors, endothelial proliferation and apoptosis (**Figure 3**) [24–26].

5. Shear stress induced Nrf2

Nrf2, a basic leucine zipper (bZIP) transcription factor is widely expressed and can be found in many organs and tissues such as the kidney, muscle, lung, heart, liver and brain [25]. The CNC family of proteins regulates gene expression, tissue differentiation and development in a variety of organs. Nrf2, perhaps the most studied of the CNC family, is responsible for the expression of phase II enzymes and a number of endogenous antioxidants including ARE-mediated processes that induce the activation of antioxidative enzymes and detoxifying enzymes, including heme oxygenase 1 (HO-1), quinone oxidoreductase (NQO1), nicotinamide adenine dinucleotide phosphate (NAD(P)H), and glutathione-S-transferase (GST) [26, 27].

Under basal conditions, the amount of Nrf2 is low due to its continuous sequestration by KEAP1 and subsequent proteasomal degradation. In this homeostatic state, Nrf2 is continuously ubiquitinated and targeted for proteasomal degradation by Kelch-like (ECH)-associated protein 1 (KEAP1). Electrophiles from endogenous and exogenous sources or other small molecules which can activate Nrf2 are thus able to do so by inactivating KEAP1 or by disrupting the KEAP1-Nrf2 binding interface [17]. Shear stress generated KLF2 induces nuclear translocation of the Nrf2 which leads to more Nrf2-ARE interactions and production of antioxidant agents (**Figure 4**).

The transcription factor that functions as the key controller of the redox homeo static gene network, Nrf2 has roles in metabolic reprogramming, proteostasis, autophagy, unfolded protein response, mitochondrial biogenesis. Inflammation, and immunity. Through this complex regulatory network, Nrf2 appears to function as a truly pleiotropic transcription factor [28, 29].

Together, more than 500 Nrf2 target genes have varying roles in mounting cellular defenses through encoding a large network of proteins, some of which catalyze phase I, II and III cytoprotective detoxification, while others are antioxidant and anti-inflammatory agents [30]. Nrf2 plays a large role in controlling cellular redox homeostasis through the regulation of key enzymes and proteins involved in synthesis, utilization and regeneration of glutathione (GSH), thioredoxin (TXN), peroxiredoxin and NADPH [31]. The activities of Nrf2 are a major determinant of the cellular redox state.

Under basal conditions, the amount of Nrf2 is low due to its continuous sequestration by KEAP1 and subsequent proteasomal degradation. In homeostatic conditions, Nrf2 is continuously ubiquitinated and targeted for proteasomal degradation by Kelch-like (ECH)-associated protein 1 (KEAP1). Electrophiles from endogenous and exogenous sources or other small molecules which can activate Nrf2 are thus able to do so by inactivating KEAP1 or by disrupting the KEAP1-Nrf2 binding interface [32]. Shear stress generated KLF2 induces nuclear translocation of the Nrf2 which leads to more Nrf2-ARE interactions and production of antioxidant agents.

Exposure to toxicants or ROS from oncogenic signaling, genetic mutations, chronic wounds, autophagy disruption, or metabolic alterations disrupt the KEAP1-Nrf2 complex leading to proteasomal degradation of Keap1 and the translocation and activation of Nrf2. KLF2 substantially enhances antioxidant activity of Nrf2 by inducing its nuclear localization and activation [16]. Nrf2 translocates into the nucleus where it heterodimerizes with a small musculoaponeurotic fibrosarcoma (sMAF) protein and binds to the antioxidant response elements (ARE), transcription

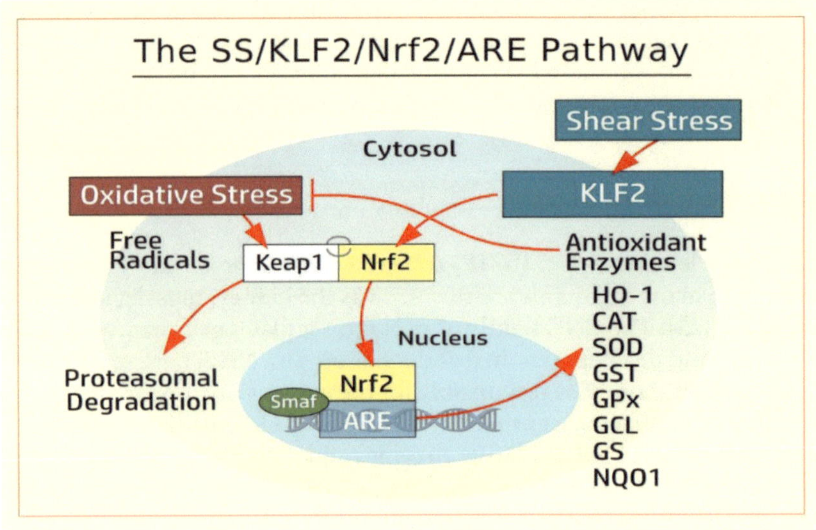

Figure 4.
The SS/KLF2/Nrf2/ARE pathway*. Shear stress upregulates KLF2 which induces the nuclear translocation and activation of Nrf2, while, concurrently, free radicals break Keap1-Nrf2 binding and lead to its degradation. Nrf2 interacts with ARE to upregulate a broad array of antioxidant elements.*

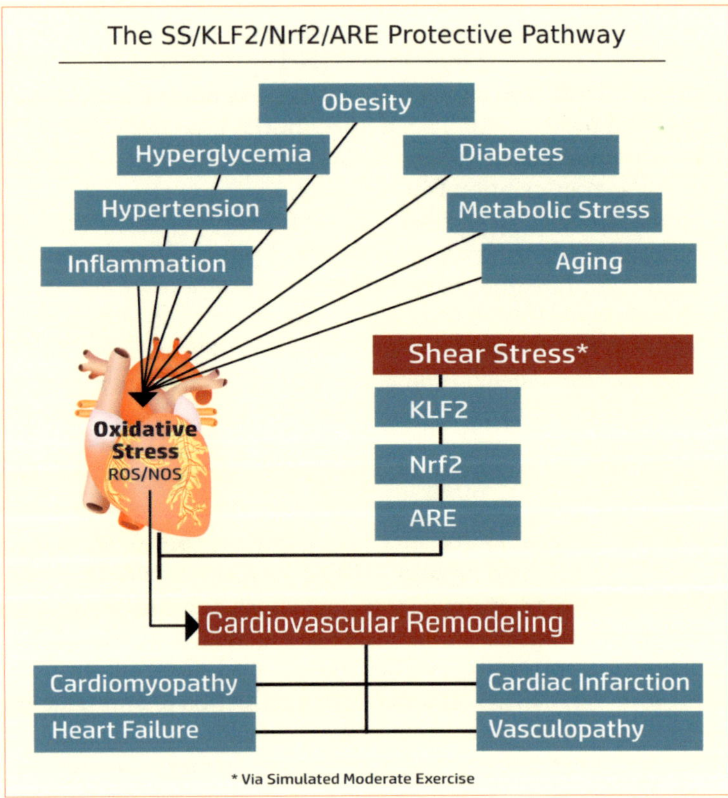

Figure 5.
The SS/KLF2/Nrf2/ARE cardioprotective pathway*. Showing the effects of the SS/Nrf2/LKF2/ARE pathway in cardiovascular health. The chart will be very similar for all diseases of oxidative stress.*

factors and cofactors to regulate its target genes, encoding proteins involved in antioxidants, detoxification, metabolism, and inflammation (**Figure 5**) [8, 17].

Since Nrf2 helps protect cells from oxidative damage, it aids in preventing major diseases. Several reports have shown the importance of the Keap1–Nrf2 system as a therapeutic target for many neurodegenerative diseases and even cancer. As a consequence, academia and the pharmaceutical industry have been investigating the Keap1–Nrf2 system attempting to increase Nrf2. Several pharmacolgical inhibitors of Keap1 have been developed to boost the effectiveness of the Nrf2/ARE pathway.

Several studies have documented the age-related decline of Nrf2. On the other hand, Narashimhan and Rajasekaran, [33] as well as Grounder and colleagues [34] found that simulated exercise in murine examples using electrical stimulation resulted in significant improvement in Nrf2 levels. In the case of Grounder's group, after six weeks of simulated moderate exercise, the aged group improved their Nrf2 levels to nearly equal those of the young group. It would seem, in light of this, that age is not the problem: lack of hemodynamic flow is the problem.

6. Shear stress generated KLF2

KLF2 is induced by SS and for more than twenty years SS importance in endothelial medicine has been steadily growing. Initially, there were investigations into whether KLF2 was an essential regulator of endothelial and organ system survival. Investigators demonstrated that KLF2 expression is increased during laminar flow in homeostasis and is reduced because of low or turbulent flow or cytokine storm. KLF2 promotes EC health through a profile of >1,000 target genes and suppresses inflammation in part through its competition with NFkB for critical transcriptional co-factors (**Figure 6**). KLF2 also promotes transcription of anti-thrombotic

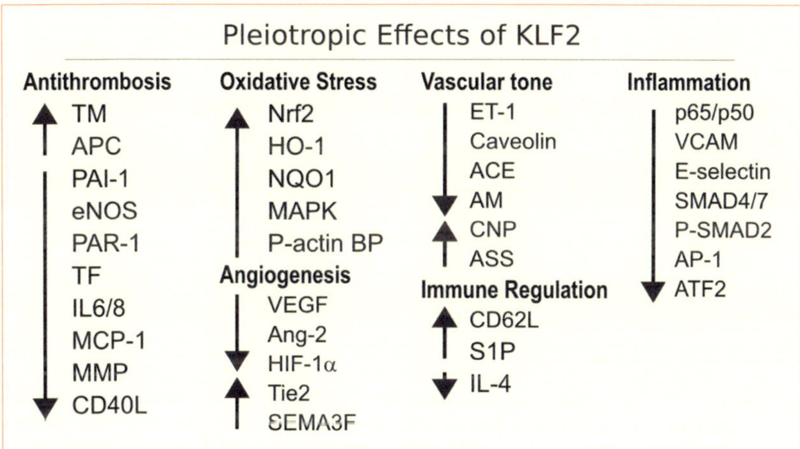

Figure 6.
Pleiotropic effects of KLF2. *CNP, C-natriuretic peptide; ET-1, endothelin-1; ASS, arginosuccinate synthase; AM, adrenomedullin; ACE, angiotensin converting enzyme; TM, thrombomodulin; APC, activated protein C; PAI-1, plasminogen activator inhibitor-1; eNOS, endothelial nitric oxide synthase; PAR-1, protease-activated receptor 1; TF, tissue factor; CD40L, CD40 ligand; MMP, matrix metalloproteinase, MCP-1, monocyte chemotactic protein 1; IL-6=8, interleukin 6=8; CD62L, CD62 ligand; S1P, sphingosine-1 phosphate; IL-4, interleukin-4; Nrf2, nuclearfactor erythroid 2-like; NQO1, NAD(P)H: quinine oxidoreductase-1; HO-1, heme oxygenease-1; MAPK, mitogen-activated protein kinase; P-actin BP, phosphorylated actin binding protein; VCAM, vascular cell adhesion molecule;ATF2, activating transcription factor 2; AP-1, activator protein 1; SMAD, Sma and Mad related protein; Ang, angiopoetin; SEMA3F, semaphorin 3F; HIF-1a, hypoxia-inducible factor 1 alpha [35].*

genes, further lending to its vasoprotective role. Endothelial KLF2 acts as a master controller promoting EC quiescence and integrity through its effects in multiple transcriptional networks [36]. In physiological conditions, the vascular endothelium is largely maintained in a quiescent and impermeable state by the constitutive activity of KLFs and the mechanosensory proteins VE-cadherin and platelet endothelial cell adhesion molecule-1 (PECAM-1). Upregulation of KLF2 results in the upregulation of Nrf-2 and eNOS together with concomitant inhibition of mitochondrial ROS production while inhibiting the transcriptional activity of NF-κB,

C-type natriuretic peptide (CNP), an autocrine and paracrine mediator is potently induced by KLF2 in cardiomyocytes and fibroblasts. It regulates a number of vital physiological functions in the cardiovascular system [37]. Circulating biomarkers of healthy endothelial function would be useful to detect the earliest deficiencies in endothelial function. CNP is an endothelial paracrine factor that has been implicated in endothelial-dependent vasodilation in certain vascular beds, in addition to suppressing neointimal hyperplasia. CNP acts on adjacent vascular smooth muscle cells by impinging on the cyclic guanosine monophosphate (cGMP) pathway that is also responsive to NO. Parmar and colleagues demonstrated that CNP is induced by statins in a KLF2-dependent manner which indicates that this molecule could be a possible biomarker of EC [38].

7. Other shear stress- generated antioxidants

The role of hemodynamics in oxidative homeostasis is of major importance. The SS/KLF2/Nrf2/ARE pathway is powerful and important, but hemodynamically driven SS produces additional synergistically acting and interacting antioxidants that can help restore redox balance. The body's endogenous antioxidant defense system relies on a complex group of enzymatic and nonenzymatic antioxidants that act against free radicals to blunt or block their pathological effects. The hemodynamically regulated antioxidants discussed below, including superoxide dismutase (SOD), catalase (CAT), glutathione peroxidase (GPX), heme oxygenase-1 (HO1), thioredoxin (TRX), and sirtuin (SIRT1), are indispensable in the defense against oxidative stress. These antioxidants are in continuous action to balance against ROS which are continuously generated in normal body metabolism, in particular through the mitochondrial energy production pathway.

7.1 Superoxide dismutase (SOD)

The first detoxification enzyme and the most powerful antioxidant in the body is SOD. It is an antioxidant enzyme that acts as the body's first line of defense against ROS. SOD catalyzes the dismutation of two molecules of superoxide anion to hydrogen peroxide (H2O2) and molecular oxygen (O2), rendering the superoxide anion less toxic (the H2O2 is further reduced by CAT and GPx). SOD is a metalloenzyme and has a metal cofactor for its activity. Three isoforms of the enzyme are identified as 1. copper/zinc (Cu/Zn SOD), 2. manganese (Mn SOD), and 3. iron (Fe/SOD) also known as extracellular (EC/SOD) [39, 40].

7.2 Catalase (CAT)

Catalases are enzymes that can neutralize hydrogen peroxide, a ubiquitous oxidant. The enzyme catalyzes the dismutation of two molecules of hydrogen peroxide into one molecule of oxygen and one molecule of water. CAT has a very high turnover rate: one catalase enzyme can convert 40 million molecules of hydrogen peroxide to oxygen and water per second. This enzyme is necessary for survival as it prevents hydrogen peroxide from accumulating to dangerous levels. Hydrogen peroxide at high levels in the body can induce cellular damage [41, 42].

7.3 Heme oxygenase-1 (HO1)

The antioxidant effects of HO-1 consist of its ability to increase glutathione levels and to degrade heme, as well as to induce biliverdin and bilirubin, both of which have potent antioxidant properties. Biliverdin is a tetrapyrrolic, water-soluble compound formed when heme is broken down into biliverdin, carbon monoxide and iron by heme oxygenase. Biliverdin is anti-mutagenic, antioxidant, anti-inflammatory, and immunosuppressant [43, 44].

7.4 Glutathione peroxidase (GPx)

Hemodynamic shear stress strongly upregulates the GPx family of extracellular antioxidant proteins that catalyze the reduction of hydrogen peroxide and lipid hydroperoxides into water and alcohols. Sometimes, its activity depends on selenium as a cofactor, and for this reason, it is often referred to as a selenocysteine peroxidase. GPx fills a crucial role by inhibiting the lipid peroxidation process to protect cells from oxidative stress [45, 46].

7.5 Thioredoxin (TRX)

The Trx system, which includes NADPH, thioredoxin reductase (TrxR), and TRX, is an important system defending against oxidative stress through its disulfide reductase activity regulating protein dithiol/disulfide balance. The cytosolic and mitochondrial Trx systems, in which TrxRs are high molecular weight seleno-enzymes, in concert with the glutathione-glutaredoxin (Grx) system (NADPH, glutathione reductase, GSH, and Grx) control cellular redox [47, 48].

7.6 Sirtuin (SIRT1)

SIRTs are a family of nicotinamide adenine dinucleotide (NAD)-dependent histone deacetylases with the ability to deacetylate histone and nonhistone targets and is linked to cellular metabolism, the redox state and survival pathways. SIRT1 deficiency in endothelial cells (ECs), vascular smooth muscle cells and monocytes/macrophages contributes to increased oxidative stress, inflammation, foam cell formation, senescence and impaired nitric oxide production. It is well established that endogenous NO generated from eNOS plays a crucial role in maintaining vascular function and homeostasis, which facilitates vascular tone, leukocyte adhesion, smooth muscle cell proliferation and migration, and platelet aggregation. Previous studies have shown that endogenous NO serves as an anti-atherosclerotic and anti-aging factor and that SIRT1 in endothelial cells regulates NO production. SIRT1 may play a crucial role in reducing inflammation and oxidative stress. Peroxisome proliferator-activated receptor gamma coactivator 1-alpha (PGC-1α) deacetylation by SIRT1 reduces oxidative stress through expression of antioxidant enzymes, including Mn-SOD. Additionally, Forkhead box protein O3a (FOXO3a) is deacetylated by SIRT1 and translocates to the nucleus, resulting in the upregulation of other antioxidant enzymes and catalases, thereby providing protection against oxidative stress [49–51].

8. Conclusions and further perspectives

The restoration of normal hemodynamics can provide a viable solution for the debilitating diseases that result from oxidative stress (**Figure 1**). While it may seem unlikely, the body has a large, complex, and robust antioxidant system with overlapping and synergistic actions. All those systems, however, are contingent upon hemodynamic activation of endocrine, autocrine and paracrine systems to restore the enzymes, proteins, genes, and other antioxidants essential for redox homeostasis. Numerous in vivo studies indicate that oxidative damage occurs from reduced levels of antioxidant enzymes rather than increased production of ROS. The utilization of hemodynamic forces and shear stress-initiated endothelial mechanotransduction to increase antioxidant enzymes is evident in the literature and warrants further investigation. The development of means and methods that enhance normal hemodynamics is also needed.

Author details

John M. Owen and Kenneth J. Dormer*
Angieon Medical Research Institute, Tulsa, OK, USA

*Address all correspondence to: kdormer@angieonresearch.org

References

[1] Cannon WB, Spirit of the Body, 1936, New York, W.W. Norton & Co.

[2] Billman GE, Homeostasis: The Underappreciated and Far Too Often Ignored Central Organizing Principle of Physiology, Front Physiol. 2020 Mar 10;11:200

[3] Dai G, Vaughn S, Zhang Y, Wang ET, Garcia-Cardena G, Gimbrone MA, Biomechanical forces in atherosclerosis-resistant vascular regions regulate endothelial redox balance via phosphoinositol 3-kinase/Akt-dependent activation of Nrf2, Circ Res 2007 Sep 28;101(7):723-33

[4] Davies PF, Flow-mediated endothelial mechanotransduction, Physiol Rev, 1995 Jul;75(3): 519-60.

[5] Endothelial Biomedicine, New York: Cambridge University Press; 2007

[6] Cahill PA, Redmond EM. Vascular endothelium - Gatekeeper of vessel health. Atherosclerosis. 2016 May; 248:97-109.

[7] Ince C, Mayeux PR, Nguyen T, Gomez H, Kellum JA, Ospina-Tascón GA, Hernandez G, Murray P, De Backer D; The Endothelium in sepsis, ADQI XIV Workgroup. Shock. 2016 Mar;45(3): 259-70.

[8] Taguchi K, Motohashi H, Yamamoto M, (2011) Molecular mechanisms of the Keap1–Nrf2 pathway in stress response and cancer evolution. Genes Cells 16(2):123-140

[9] Aird WC. Spatial and temporal dynamics of the endothelium. J Thromb Haemost. 2005 Jul;3(7):1392-406.

[10] Gimbrone MA, The Gordon Wilson lecture. Understanding vascular endothelium: a pilgrim's progress. Endothelial dysfunction, biomechanical forces and the pathobiology of atherosclerosis, Trans Am Clin Climatol Assoc, 2010;121:115-27

[11] Malek AM, Alper SI, Izumo S, Hemodynamic Shear Stress and Its Role in Atherosclerosis. JAMA 12;99; vol 282, no 21

[12] Hill MA, Meininger GA. Arteriolar vascular smooth muscle cells: mechanotransducers in a complex environment. Int J Biochem Cell Biol. 2012 Sep;44(9):1505-10.

[13] Zhang Y, Liao B, Li M, Cheng M, Fu Y, Liu Q, Chen Q, Liu H, Fang Y, Zhang G, Yu F. Shear stress regulates endothelial cell function through SRB1-eNOS signaling pathway. Cardiovasc Ther. 2016 Oct;34(5):308-13.

[14] Mammoto A, Mammoto T, Ingber DE. Mechanosensitive mechanisms in transcriptional regulation. J Cell Sci. 2012 Jul 1;125 (Pt 13):3061-73.

[15] Sackner MA, Adams JA, Endothelial pulsatile shear stress is a backstop for COVID-19, Emerging Topics in Life Sciences (2020) 4 391-399

[16] Fledderus JO, Thienen JV, Boon RA, Dekker RJ, Rohlena J, Volger OL, Bijnens AJJ, Daemen MJAP, Kuiper J, van Berkel TJC, Pannekoek H, Horrevoets AJG, Prolonged shear stress and KLF2 suppress constitutive proinflammatory transcription through inhibition ofATF2, Blood First Edition Paper, January 23, 2007;

[17] Naidu S, Dinkova-Kostova AT. 2020 KEAP1, a cysteine-based sensor and a drug target for the prevention and treatment of chronic disease. Open Biol. 10: 200105.

[18] Rahman T, Hosen I, Towhidul MM, Hossain I, Shekhar U, Oxidative stress

and human health, Advances in Bioscience and Biotechnology, 2012, 3, 997-1019

[19] Small DM, Gobe GC, Oxidative Stress and Antioxidant Therapy in Chronic Kidney and Cardiovascular Disease in Oxidative Stress and Chronic Degenerative Diseases - A Role for Antioxidants, Intechopen, London, 2013

[20] Chi Hyun Kim, Eui-Bae Jeung, Yeong-Min Yoo, Combined Fluid Shear Stress and Melatonin Enhances the ERK/Akt/mTOR Signal in Cilia-Less MC3T3-E1 Preosteoblast Cells, Int. J. Mol. Sci. 2018, 19, 2929;

[21] Wassermana SM, Topper JN, Adaptation of the endothelium to fuid flow: in vitro analyses of gene expression and in vivo implications, Vascular Medicine 2004; 9: 35¡45

[22] Ursini F, Maiorino M, Forman HJ, Redox homeostasis: The Golden Mean of healthy living, Redox Biology 8(2016) 205-215

[23] Chong ZZ, Shang YC,Wang S, Maiese K, SIRT1: New avenues of discovery for disorders of oxidative stress, Expert Opin Ther Targets. 2012 February; 16(2): 167-178.

[24] Krüger-Genge A, Blocki A, Franke R, Jung F, Vascular Endothelial Cell Biology: An Update, Int J Mol Sci, 2019 Sep 7;20(18):4411.

[25] Paszkowiak JJ, Dardik A, Arterial Wall Shear Stress: Observations from the Bench to the Bedside, Vasc Endovascular Surg, Jan-Feb 2003;37 (1):47-57

[26] Schneider KS, Chan JY, Emerging role of Nrf2 in adipocytes and adipose biology. Adv Nutr. 2013 Jan 1;4(1):62-6.

[27] Deshmukh P, Unni S, Krishnappa G, Padmanabhan B, The Keap1–Nrf2 pathway: promising therapeutic target to counteract ROS-mediated damage in cancers and neurodegenerative diseases, Biophys Rev (2017) 9:41-56

[28] Dong J, Zhang X, Wang S, Xu C, Gao M, Liu S, Li X, Cheng N, Han Y, Wang X and Han Y, Thymoquinone Prevents Dopaminergic Neurodegeneration by Attenuating Oxidative Stress Via the Nrf2/ARE Pathway. Front. Pharmacol. 2021 11:615598.

[29] Uryash A, Bassuk J, Kurlansky P, Altamirano F, Lopez JR, Adams JA (2015) Antioxidant Properties of Whole Body Periodic Acceleration (pGz). PLoS ONE 10(7): e0131392

[30] Feng He, Xiaoli Ru, Tao Wen, Nrf2, a Transcription Factor for Stress Response and Beyond Int J Mol Sci. 2020 Jul; 21(13): 4777

[31] Chen, Maltagliati AJ, Nrf2 at the heart of oxidative stress and cardiac protection, Physiol Genomics. 2018 Feb 1; 50(2): 77-97.

[32] Tebaya LE, Robertsona H, Durant ST, Vitalec SR, Trevor M. Penning TM, Dinkova-Kostovaa AT, Hayes JD, Mechanisms of activation of the transcription factor Nrf2 by redox stressors, nutrient cues, and energy status and the pathways through which it attenuates degenerative disease, Free Radic Biol Med. 2015 November; 88(0 0): 108-146

[33] Taguchi K, Kensler TW, Nrf2 in Liver Toxicology, Arch Pharm Res. 2020 Mar; 43(3): 337-349

[34] Narasimhan M, Rajasekaran NS (2016) Exercise, Nrf2 and Antioxidant Signaling in Cardiac Aging. Front. Physiol. 7:241.

[35] Parmar KM, Nambudiri V, Dai G, Larman HB, Gimbrone, MA, Jr., and Garcıa-Cardena G, Statins Exert Endothelial Atheroprotective Effects via

the KLF2 Transcription Factor, Jour Bio Chem, Vol. 280, No. 29, Issue of July 22, pp. 26714-26719, 2005

[36] Gounder SS, Kannan S, Devadoss D, Miller CJ, Whitehead KS, et al, Impaired Transcriptional Activity of Nrf2 in Age-Related Myocardial Oxidative Stress Is Reversible by Moderate Exercise Training. PLoS ONE 7(9): e45697.

[37] Sweet DR, Lam C and Jain MK, Evolutionary Protection of Krüppel-Like Factors 2 and 4 in the Development of the Mature Hemovascular System. Front. Cardiovasc. Med. 8:645719.

[38] Moyes AJ, Hobbs AJ, C-Type Natriuretic Peptide: A Multifaceted Paracrine Regulator in the Heart and Vasculature, Int. J. Mol. Sci. 2019, 20, 2281;

[39] Nayak L, Lin Z, Jain MK, Go with the Flow: How Kruppel-Like Factor 2 Regulates the Vasoprotective Effects of Shear Stress, Antiox & Redox Sig, Vol 15, No 5, 2011

[40] Ighodaro OM, Akinloye OA, First line defense antioxidants-superoxide dismutase (SOD), catalase (CAT) and glutathione peroxidase (GPX): Their fundamental role in the entire antioxidant defence grid, Alexandria Journal of Medicine 54 (2018) 287-293

[41] Emina ČolakLepša Žorić, in Handbook of Nutrition, Diet, and the Eye (Second Edition), 2019, Academic Press, Salt Lake City

[42] J. Everse, in Encyclopedia of Biological Chemistry (Second Edition), 2013, Academic Press, Salt Lake City

[43] Su C, Li M, Catalase Kinetics (internet) found at: web.mit.edu/chrissu/Public/5310lab3.pdf

[44] Turkseven S, Kruger A, Mingone, CJ, Pawel Kaminski P, Rodella IF, Wolin MS, Abraham NG, Antioxidant mechanism of heme oxygenase-1 involves an increase in superoxide dismutase and catalase in experimental diabetes, Am J Physiol Heart Circ Physiol 289: H701–H707, 2005.

[45] Maeve C. McDonnell, Shamim S. Mohiuddin, Biochemistry, Biliverdin In: StatPearls, Treasure Island FL: StatPearls Publishing; 2021

[46] Ottaviano FG, Tang S, Handy DE, Loscalzo J, Regulation of the extracellular antioxidant selenoprotein plasma glutathione peroxidase (GPx-3) in mammalian cells, Mol Cell Biochem, 2009 Jul;327(1-2):111-26.

[47] Uryash A, Bassuk J, Kurlansky P, Altamirano F, Lopez JR, Adams JA, Antioxidant Properties of Whole Body Periodic Acceleration (pGz). PLoS One 10(7): e0131392.

[48] Hideyuki Yamawaki, Shi Pan, Richard T Lee, Bradford C Berk Fluid shear stress inhibits vascular inflammation by decreasing thioredoxin-interacting protein in endothelial cells. 2005 Mar;115(3):733-8.

[49] Jun Lu, Arne Holmgren, The thioredoxin antioxidant system, Free Radic Biol Med, 014 Jan;66:75-87.

[50] Philp A, Schenk S, Unraveling the complexities of SIRT1-mediated mitochondrial regulation in skeletal muscle, Exerc Sport Sci Rev. 2013 July ; 41(3): 174-181.

[51] Kitada M, Ogura Y, Koya D. The protective role of Sirt1 in vascular tissue: its relationship to vascular aging and atherosclerosis. Aging (Albany NY). 2016 Oct 15;8(10):2290-2307.

Section 3

Thalassemia

Chapter 7

An Early Diagnosis of Thalassemia: A Boon to a Healthy Society

Nitu Nigam, Prithvi Kumar Singh, Suhasini Bhatnagar, Sanjay Kumar Nigam and Anil Kumar Tripathi

Abstract

The β-thalassemia is a hereditary blood disorders, characterized by reduced or absent synthesis of the hemoglobin beta chain that cause microcytic hypochromic anemia. An early diagnosis, economical test, awareness programs and prenatal screening will be a milestone for the eradication of this genetic disorder and to reduce burden of the health sector of a country subsequently the economics. Initially, the diagnosis of β-thalassemia depends on the hematological tests with red cell indices that disclosed the microcytic hypochromic anemia. Hemoglobin analysis shows the abnormal peripheral blood smear with nucleated red blood cells, and reduced amounts of hemoglobin A (HbA). In severe anemia, the hemoglobin analysis by HPLC reveals decreased quantities of HbA and increased the level of hemoglobin F (HbF). The decrease level of MCV and MCH are also associated with β-thalassemia. There are various different molecular techniques such as ARMS PCR, allele-specific PCR, Gap PCR, denaturing gradient gel electrophoresis, reverse dot blotting, DGGE, SSCP, HRM, MLPA, sequencing technology and microarray available to identify the globin chain gene mutations. These molecular techniques can be clustered for detection by mutation types and alteration in gene sequences.

Keywords: β-thalassemia, Microcytic Hypochromic Anemia, HPLC, Mutation, Molecular techniques

1. Introduction

There are many tests available for the diagnosis of thalassemia and hemoglobinopathies. However, diagnosis of these conditions that is sufficiently accurate for most of the clinical conditions can usually be established from complete family history (pedigree analysis) and a complete clinical and hematological examination of the patient and their family members.

Generally, doctors everywhere in the world diagnose thalassemia using blood tests which include a complete blood count and special tests for hemoglobin abnormalities. Initially, the primary screening of the thalassemia depends on the complete blood count (CBC), detection of carriers is done by hematological tests with red cell indices and microcytic hypochromic with mild anemia. The high performance liquid chromatography (HPLC) or capillary zone electrophoresis is used for the identification of qualitative and quantitative assessment of hemoglobin (Hb) components. HbA2 test is the most significant identification test for β thalassemia minor, but it can be vary by the presence of defective δ-thalassemia. In earlier days various molecular techniques e.g. amplification refractory mutation specific-polymerase chain reaction ((ARMS-PCR), GAP PCR have been used for the detection of mutations in β- and α-thalassemia, which may help in the prenatal diagnosis of thalassemia hemoglobinopathy in a limited time. Recently the evolution of Next- Generation Sequencing (NGS) has taken an important place for the diagnosis of thalassemia as a confirmatory diagnostic test. The NGS has been introduced for the characterization of both α- and β-thalassemia genes. It gives an accurate diagnosis of thalassemia, although NGS predicts much higher carrier frequencies. The molecular analysis is fundamental to foresee the severe blood transfusion-dependent thalassemia cases to the mild or no blood transfusion. The prenatal diagnosis based on DNA by amniocentesis and chorionic villus sampling (CVS) is equivalent required to detect the genetic abnormalities of the foetus as per expertise. Now a day's NIPT (Noninvasive prenatal testing) is a technique to identify the genetic abnormalities of the foetus. This testing examines the small remains of foetus DNA that are circulating in a pregnant mother's blood. An appropriate lab conclusion is urgent for describing the various types of thalassemia hemoglobinopathy with a significant association for prevention and treatment.

This chapter will explain all of these tests, the information of which will be useful for those who are working and interested in the diagnosis of thalassemia hemoglobinopathy.

2. Diagnostic strategies for thalassemia

In many diagnostic labs, the diagnostic strategies have been established for the diagnosis of thalassemia from the simple PCR to NGS for the detection of common, less common and rare thalassemia mutations [1]. Although there are lots of PCR technologies available in the laboratory but most of the diagnostics labs are using the simple and robust technique on allele- specific oligonucleotide hybridization or allele-specific priming, e.g. reverse dot-blotting or ARMS-PCR for the identification of the beta-thalassemia carrier [2]. This approach helps in identifying the common and less common mutations in 90% of cases. The rare mutations will be identifying by secondary screening. The mutation which remains unidentified after these two screenings will be then characterized by DNA sequencing [3]. DNA sequencing is a technique used to identify the specific arrangement of nucleotide bases (A, C, G, and T) in a DNA. The DNA carries the information a cell needs to collect protein and RNA molecule. DNA grouping data is imperative to researchers examining the elements of qualities. DNA sequence information is mandatory to researchers examining the functions of genes [4].

However worldwide laboratories, it has been noticed that a result of the migration of the natives with different ethnicity has led to increasing the variety of hemoglobinopathy and thalassemia mutations that need to be identified. Because of migration the genetic makeup of the population has mixed, these Hb variations are presently seen everywhere. Thus, the compound heterozygous conditions (Hb D–β thalassemia, Hb E–β thalassemia and Hb Sβ thalassemia) are seen in various places [5]. Molecular indicative research centers in such nations should have the specialized skill, equipment, and diagnostic approach to distinguish an enormous variety of mutations rapidly for prenatal diagnosis, and these labs use DNA sequencing as the principle evaluating strategy for the diagnosis of β-thalassemia point mutations.

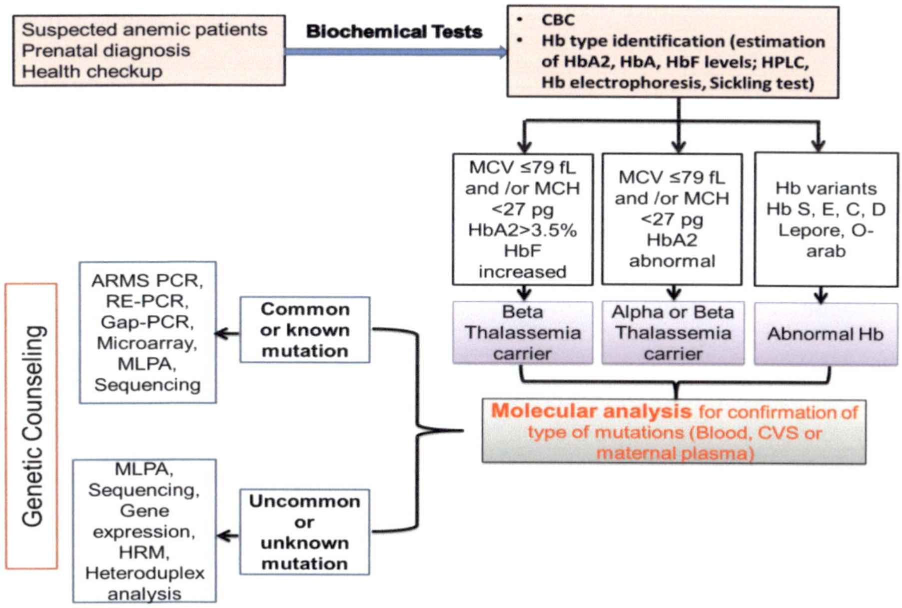

Figure 1.
Flow chart explain the diagnosis of thalassemia from primary to secondary screening.

The **flow chart** shows the step wise diagnostic model for thalassemia (**Figure 1**). The hematological and biochemical tests such as complete blood count (CBC) and Hb type identification were performed for suspected anemic patients. The altered or abnormal biochemical findings will be further confirmed by molecular analysis. The Allele-specific oligonucleotide (ASO), reverse dot-blot (RDB), ARMS-PCR, RE-PCR, Gap-PCR, MLPA and sequencing will be performed for known and unknown mutations. Finally prenatal will be done followed by genetic counseling in the affected family.

The government of the country should make various screening and diagnosis mandatory for the eradication of thalassemia. They are indicated as under mentioned [6]. The country Greece has followed this rule now Greece is a thalassemia free country.

- Premarriage screening
- Antenatal screening
- Preconception screening
- Neonatal screening
- Preoperative/pre-anesthesia screening
- Genetic counseling

2.1 Premarriage screening

Premarriage screening should be implemented to detect β thalassemia carriers and hemoglobinopathies such as sickle cell trait. In developing country, it is not frequently acceptable because of social stigma and reasons in the general public.

The premarriage screening should be possible in universities and colleges, schools, or public places such as theater, shopping mall etc. Earlier it was performed in where the predominance of thalassemia is high. But nowadays due to the mixing up of the gene pool this screening is recommended in all colleges where students are of marriageable age. Our lab has performed this screening in the Tharu tribal area of the Kheri Lakhimpur, Uttar Pradesh, India. We have collected more than 700 samples from the college and carrier screening of sickle cell disease has been performed. It is a belt of HBS [7].

2.2 Antenatal screening

In antenatal screening irrespective of gestational age of all pregnant women should be screened for carrier status of thalassemia and hemoglobinopathies. The spouse of the affected female should be test for mutation such as α-, β thalassemia and hemoglobinopathies (HbS trait, Hb E trait, Hb D trait etc.). The Prenatal diagnosis have to be encouraged if the foetus is in danger for the having thalassemia mutations e.g. α-, β thalassemia and hemoglobinopathies (HbS trait, Hb E trait, Hb D trait etc.). If the couple found positive for thalassemia during antenatal screening they can decide for prenatal diagnosis and subsequent pregnancy [8–11].

2.3 Preconception screening

While troublesome circumstances are in the developing country like India, this ought to be done but mostly females frequently do not enroll in antenatal centres before 12 weeks of gestation. A similar methodology with respect to antenatal screening ought to be followed. The preconception screening is important for all couples coming to IVF in infertility clinics. If the female is a thalassemia carrier then her husband or sperm donor should be screened and vice versa [8–11].

2.4 Neonatal screening

Infant screening is mainly suggested in haemoglobinopathies such as sickle cell diseases, prevalent in tribal and urban populations. If possible, neonatal screening has to be implemented universally where infants are at high-risk for homozygous β-thalassemia and all instances of HbS–β thalassemia. This methodology will miss a couple of instances of sickle β-thalassemia when the mother is a β-thalassemia carrier and the father is a carrier of HbS. All babies with significant hemoglobinopathy should be re-tested using molecular technology to confirm the diagnosis within three months after birth [8–11].

2.5 Preoperative/pre-anesthesia screening

Preoperative/pre-anesthesia screening of patients is preventive measure where the prevalence of HbS is high. As the presence of sickle hemoglobin might interfere during preoperative and postoperative procedures [8–11].

2.6 Genetic counseling

Genetic counseling should be given by an expert to affected thalassemia families. The advice can likewise be given by a trained genetic counselor, a hematologist, or a pediatrician [8–11].

3. Screening methods for thalassemia

3.1 Hematological screening

The primary screening of thalassemia is based on (MCV, MCH) values, Hb A2 levels, and F levels by automated blood analyzer, Hb electrophoresis and HPLC, respectively. The secondary screening encompasses further hematological studies to detect suspected variant such as sickle solubility, iron levels and heat instability tests. The **complete blood count** (CBC) measures the exact amount of hemoglobin and different types of blood cells. As it is clear that when there are mutations in the genes which are responsible for the hemoglobin synthesis that can lead to hemoglobinopathies. It can be further categorized based on of the type of mutations on the globin genes (α, β, δβ) and abnormal structural variants such as Hb Lepore and Hb E. These discrepancies in the mutations on the globin chains led to the different phenotypes of thalassemia. When there is a substantial increase of HbF in adults caused by the γ-globin chains synthesis after birth without any important clinical or hematological manifestations, this condition is known as Hereditary Persistence of Foetal Hemoglobin (HPFH) [12].

The variations from the typical hematological phenotypes of β-thalassemia carrier include:

- Decrease level of MCV and MCH with marginal or normal levels of Hb A2 when person must consider α -thalassemia, heterozygosity for mild β-thalassemia mutations, iron deficiency, heterozygosity for eγδβ-thalassemia.

- Normal/Borderline MCV and MCH levels with higher Hb A2 when person must deliberate co-inheritance of alpha and beta-thalassemia.

- Normal Hb A2 with Normal or decreased red cell indices but raised Hb F level when person should consider heterozygous δβ -thalassemia or HPFH.

- The deranged hematological and biochemical value after primary and secondary screening is confirmed by molecular analysis.

Red cell distribution width (RDW) test is an estimation of the range in the volume and size of the RBC (Red blood cell, Erythrocytes). The RBC move oxygen from the lungs to each cell in the body which helps in normal development. If the RBC's are bigger than the normal size, it could show a health problem. The RDW test is normally used to investigate anemia. It is a condition wherein the RBC cannot carry sufficient oxygen to the body. The RDW test may likewise be utilized to diagnose thalassemia [13]. This change in RDW is very remarkable in thalassemia from HbH sickness to thalassemia minor [14]. In thalassemia, RBC count is increased with microcytic anemia in compared to iron deficiency anemia (IDA) and iron deficiency where RBC count is relatively decreased. That's why in any hemoglobin disorder RBC count and RDW level should not be considered as only evaluating methods.

3.2 Biochemical screening

As per the International Committee for Standardization in Hematology (ICSH) in 1978, has recommended three kinds of lab tests for the diagnosis of thalassemia and hemoglobinopathies [15]. In that rule, the screening research facility ought to have the option to perform the alkaline electrophoresis. The reference lab approved by ICSH needed to perform challenging tests like globin electrophoresis and citrate agar electrophoresis. It is necessary to have manual strides all through hemoglobin investigation from reagent availability, electrophoresis, and information examination, and accordingly, the experience of the research centre proficient was a key to fruitful recognizable proof. Late improvement of lab strategies and expanded information on hemoglobinopathy and thalassemia has determined the distribution of refreshed rules [16]. The British Committee for Standards in Hematology suggests possible recognizable proof of hemoglobins on at least two procedures and gets conclusive ID as that dependent on DNA examination, protein sequencing or mass spectrometry.

Electrophoresis is a strategy used to isolate atoms or mixtures dependent on their movement design in a gel and electrical field. It is more commonly utilized in diagnostics labs for protein electrophoresis and the separation of some isoenzymes. Manual planning of gel and electrophoresis is infrequently utilized in evolved nations as further developed and computerized methods, for example, narrow electrophoresis are accessible.

In early long stretches of finding cellulose acetic acid derivation electrophoresis is an agent custom electrophoresis strategy. It gives distinguishing proof of Hb A, F, S/G/D, C/E, and H and different variations [16].

High Performance fluid chromatography (HPLC) is a technique in which isolate mixtures of atoms dependent on their substance qualities. Numerous division standards like affinity, segment and size are accessible; for hemoglobin, molecule exchange chromatography is proficient and frequently utilized. The technique can be additionally physically worked, yet as of late completely mechanized frameworks are accessible. Those frameworks might be devoted to hemoglobin examination for haemoglobinopathies and thalassemia (**Figure 2**).

HPLC is useful for the finding of β-thalassemia carrier since that HbA2 can be correctly quantitated [16]. Similar other HPLC methods, a wary control of insightful conditions like segment temperature, stream rate, and backup conditions are important.

Mass spectrometry is a procedure to recognize molecules dependent on their mass (sub-atomic weight) to charge proportion. In this procedure, the molecule of the interest required minimal specific binding reagents. The simple analytical norm allows less interfering but rather more precise detection. In this technique, the examination of hemoglobin is not easy as it required specialized ability for the investigation of proteins and costly instruments. Other than detecting of hemoglobin based on the intact molecular weight, it can likewise examine the amino acids sequence. Mass spectrometry is helpful tool for the detection of new confirmation and DNA sequencing variations [17].

3.3 Advance molecular techniques for the characterization of thalassemia

There are various different molecular techniques available to identify the globin chain gene mutations. These molecular techniques can be clustered for detection by mutation types such as structural variations (Gene deletion, duplication, or triplication) and alteration in gene sequences (Insertion, substitution, or short insertion/deletions) [18, 19]. Over 90% of α-thalassemia patients are caused by gene deletion. Approximate 10% of the α-thalassemia cases are due to the alteration in the gene sequence such as single nucleotide insertion, deletion or substitution [20]. The α-globin gene group encode identical protein which consists of exceptionally homologous genes as well as 2 HBA genes. The gene deletions in α-thalassemia are mainly cause due to imbalanced crossing over between these homologous regions during meiosis [21]. The most well-known deletion of 3.7 kb and 4.2 kb has been

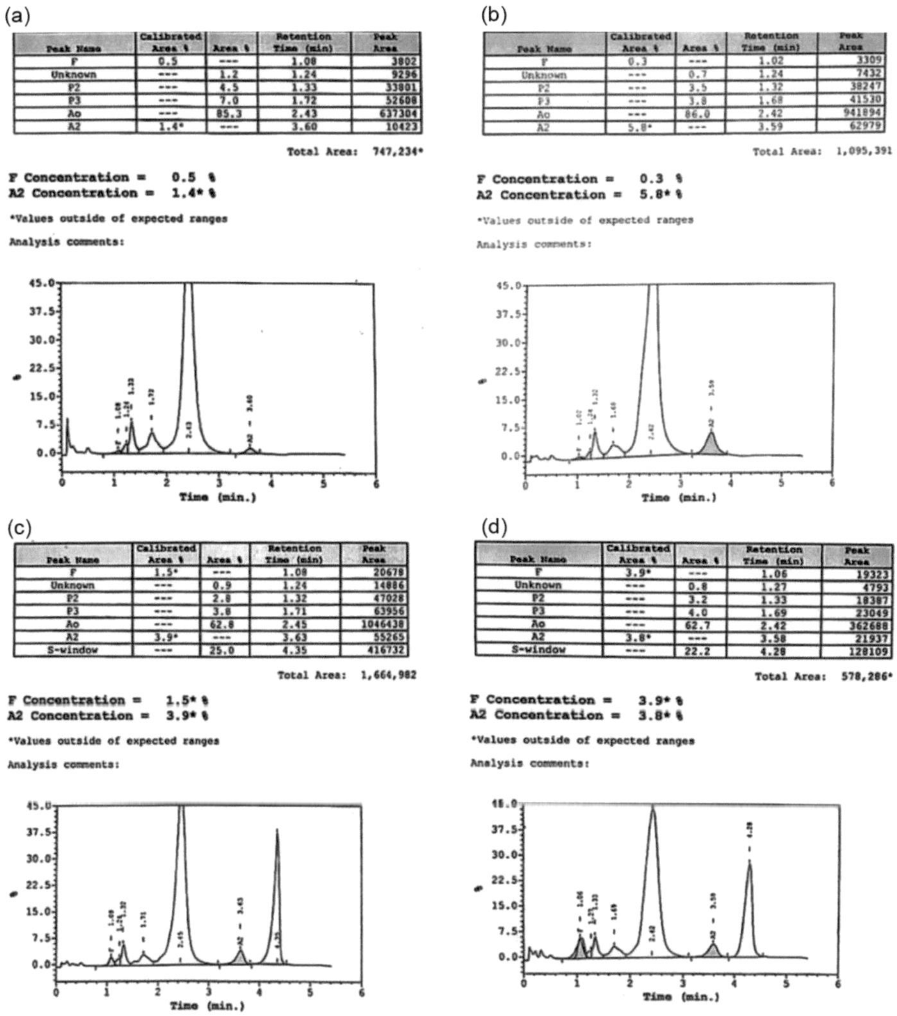

Figure 2.
HPLC of various types of haemoglobinopathies and thalassemia; (a) Normal hemoglobin (Hb); (b) β-thalassemia trait with A2 fraction on 5.8%; (c) and (d) compound heterozygous HbS and β-thalassemia trait (Nigam et al., 2020).

reported [22]. More than 90% of β-thalassemia cases are caused by the alteration in the gene sequence as compared to α-thalassemia. Approximately, 280 gene sequence alterations are related with β-thalassemia in which some mutations are caused by the deletion of gene including the HBB gene [22].

Gap polymerase chain reaction (PCR) technique is mainly used for the detection of deletion. The southern blotting may be used for unknown deletions with the help of using probes. The MLPA (multiplex ligation-dependent probe amplification) technique can identify both known and unknown deletions. It is commonly used in diagnostics labs for its highly sensitive and is easy to use. Common mutation or alteration in gene sequences can be detected by using techniques such as amplification refractory mutation specific (ARMS) PCR, allele-specific PCR, denaturing gradient gel electrophoresis, reverse dot blotting, DGGE, SSCP, HRM (High resolution melting), sequencing technology and microarray in a cost-effective manner.

Allele-specific oligonucleotide (ASO) hybridization and reverse dot-blot (RDB) techniques are used for the detection of known mutations. In this technique, the PCR products (amplified target DNA sequences) are hybridize with two oligonucleotide probes; one complementary to mutant sequence and the other to a normal sequence. The normal probes hybridized with normal individuals [23, 24]. The ASO's vary from one another by the changes in a single nucleotide. The analysis of the change in a single nucleotide in DNA using hybridization with ASO probes that have been bound to a nylon membrane in the form of the dot after restriction endonuclease digestion and electrophoresis [23, 25–26]. The ASO probes are specific and complementary for the several alleles, to detect known mutation or single nucleotide polymorphism. This dot blot technique is used for one or two major mutations [27]. This was surprised by the development of the reverse dot-blotting technique, in which, the panel of mutation-specific probes is hybridized with amplified DNA, fixed to a nylon membrane. This method is viable with the ideal procedure for screening β-thalassemia mutations, using a panel of common known mutations for the primary screening and a panel of uncommon for the second screening [28].

Primer Specific Amplification– amplification refractory mutation specific (ARMS) PCR is the most commonly used technique for the detection of β-thalassemia mutations [29, 30]. This technique is a simple, quick screening assay; does not require the latest technology [30]. The ARMS primers have been made for the detection of common mutations of β-thalassemia [31]. In various countries like India and Pakistan, this technique quit well-known for screening and prenatal diagnosis due to its quick and low cost [32, 33]. The ARMS PCR is able to detect multiple known mutations in a single assay [29, 34, 35]. It is also identified the change in target DNA is heterozygous or homozygous. With the help of ARMS primers (common forward primer and two reverse one mutant and other to the normal primer sequence), ARMS PCR is differentiated homozygote or heterozygote.

Recently, ARMS PCR technology has been improved for the detection of both normal and altered alleles with internal positive control are detected in a single tube assay [35], referred to as tetra primer ARMS-PCR. Two pairs of primers are used in tetra, ARMS-PCR in which one pair of primers for flanking regions and other pair primers are complementary to different strands. These primers amplify the two different bases that are located in a single position of the globin gene. The different alleles (mutant and wild type) can be detected on an agarose gel based on their sizes (**Figure 3**). This ARMS PCR technique has been useful in the diagnosis of β-thalassemia mutations. Multiplex ARMS PCR can be screened for more than one mutation a single reaction by multiplexing the ARMS primers attached with a common primer [36].

Restriction enzyme PCR (RE-PCR)/restriction fragment length polymorphism (RFLP) has had a restricted diagnostic role due to few β-thalassemia mutations restriction sites. Even though its use can be enlarged by the artificial formation of a restriction that includes the target mutation [37]. The fundamental use of RE-PCR has been for the examination of β-globin gene haplotypes to define the origin of mutations in the globin gene in various ethnic groups [38].

The target genomic DNA containing the mutation is amplified by PCR with the help of specific primers. The PCR products are digested by a specific restriction enzyme. The digested PCR products are separated on the agarose gel. The digested PCR products are separated according to their molecular weight (size). Based on restriction site (presence or absence) is determine the size or pattern of PCR products. This RE analysis is simple, relatively economical, and powerful prompting unequivocal outcomes; the RE-PCR-based technique is a precious molecular diagnostic tool. Be that as it may, it is restricted in its application as just an extent of the alpha-thalassemia, β-thalassemia mutation, and hemoglobin variations, naturally generate restriction sites [26, 39].

Gap-PCR is a rapid, simple and non-radioactive technique that identified the deletions of globin gene. Gap-PCR amplifies the deleted DNA sequence in target

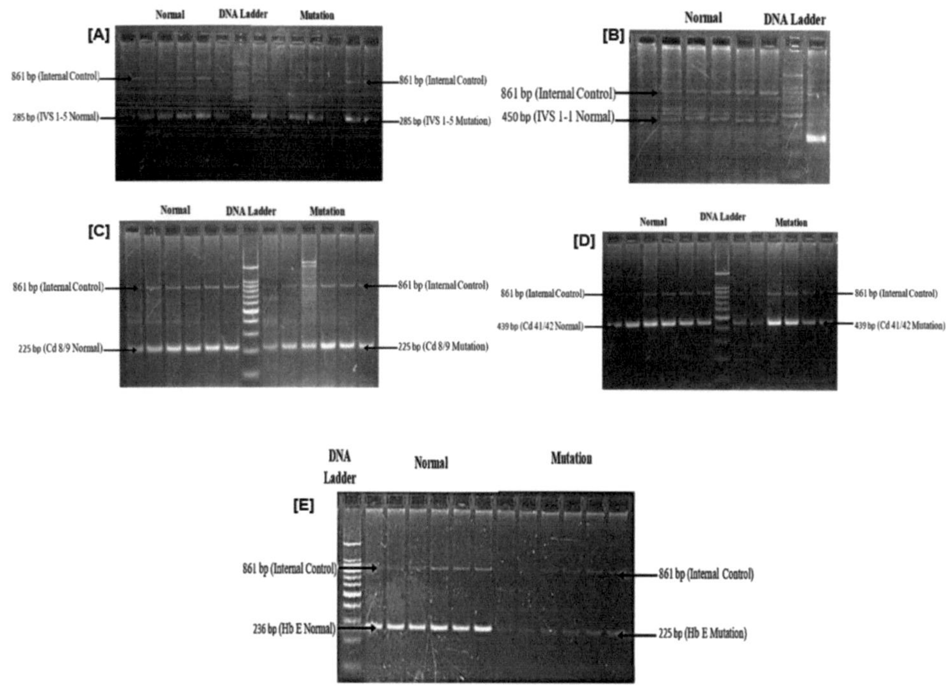

Figure 3.
Gel image shows the multiplex ARMS PCR for five common mutations: [A] IVS 1–5 mutation/normal; [B] IVS 1–5 mutation/normal; [C] Cd 8&9 mutation/normal; [D] Cd 41&42 mutation/normal; and [E] Hb E mutation/normal.

DNA by using the flanking primers for this region. This flanking primer pairs are making a unique PCR product, smaller for mutant sequence as compared to wild type sequence [40]. In alpha-thalassaemia mostly mutation are deletion types e.g. $-\alpha^{3.7}$, $-\alpha^{4.2}$ deletion, HbLepore and HPFH deletion etc. [41, 42]. In Asian Indians, 619 bp deletion is found in β-thalassaemia. The limitation of Gap-PCR technique is that the deletion endpoints must be known for primers designing.

Denaturing gradient gel electrophoresis (DGGE) and single-stranded conformation polymorphism (SSCP) are used for the screening of the unknown mutations. The DGGE technique allows the DNA fragments differing by single nucleotide base change according to its melting characteristics [43, 44]. In SSCP, single nucleotide substitutions can be identified but the efficiency is between 70 to 90%. In both techniques, the PCR products mobility on gel is altered in mutant sequence as compared to normal sequence and further confirmed by sequencing. Once the unknown mutation has been detected by DNA sequencing, these DNA can be used as a control to develop ASO and ARMS primer for its detection in further cases. The only disadvantage of these techniques is that it cannot be applied to hemoglobinopathies.

Heteroduplex analysis is another method utilizes non-denaturing gel electrophoresis. By annealing and amplified target DNA fragment with an amplified hetroduplex generator molecule, the unique heteroduplex pattern can be generated for each mutation of 130 bases in length [45].

High Resolution Melting (HRM) investigation is a moderately new, post-PCR examination technique used to recognize the changes in DNA sequences. The technique depends on distinguishing little contrasts in PCR melting (dissociation) curves. It is empowered by further developed dsDNA- binding dyes conjunction with real-time PCR instrumentation that has exact temperature ramp control and advanced data capture capabilities. Data are analyzed and manipulated using software designed specifically for HRM analysis. The HRM is an analytical platform for rapid prenatal and postnatal diagnosis of β-thalassemia common among Southeast Asian population. The advantage of HRM is a simple, economical with fast workflow platform for the diagnosis of single gene disorder [46, 47].

Multiplex Ligation-Dependent Probe Amplification (MLPA) is versatile useful technique for the diagnosis of copy number variations (CNVs) from complete chromosomes to single exons associated with genetic disorders and tumors. To detect DNA methylation changes can be detected by Methylation-specific MLPA (MS-MLPA) which is sensitive enough to distinguish the changes in disease causing genes from highly similar pseudogenes [48].

The multiplex PCR technique that amplify up to 60 probes by using one pair of primer. The PCR amplicon with novel genomic target and length are fluorescently labeled and identified by capillary electrophoresis. The number of genomic sequence of interest is determined by comparing the peak pattern with reference sample [49].

Since, the varied molecular basis of α-thalassemia and β- thalassemia mutation are uncommon and difficult to detect. In addition to well-established methods, MLPA is known as an effective, simple and unambiguous technique for the identification and classification of deletions and duplications in thalassemia [48–50].

Direct DNA sequencing is a technique used to identify the specific arrangement of nucleotide bases (A, C, G, and T) in a DNA. The DNA carries the information a cell needs to collect protein and RNA atoms. DNA grouping data is imperative to researchers examining the elements of qualities. DNA sequence information is mandatory to researchers examining the functions of genes [51]. In beta globin gene, most of mutations can be identified in two sequence reads but for alpha-globin only single read required. The advantage of the simple sequencing method for beta globin gene play in important role in prenatal diagnosis where the mother is HBS carrier and father is unavailable [52]. If the developing embryo found HBS mutation then further analysis of beta globin sequences is must to assure that foetus does not have compound heterozygous state HBS and coexisting beta-thalassaemia [53].

This is a significant result for the haemoglobinopathies as most cases include carrier testing, subsequently arrangement follows much of the time require checking by eye. There are two diverse sequencing sciences accessible dependent on the Sanger technique [54–56], dye primer and dye eliminator; they differ from each other in the way wherein the fluorescent level is incorporated during linear cycling. Despite the fact that the dye eliminator is more straightforward to set-up the signal from each nucleotide is less dependable making the dye primer chemistry more suitable for heterozygote detection and it is friendlier with sequence analysis programming. The dye eliminator would have more application in X- linked diseases, for instance, G6PD transformations in which influenced males are hemizygous and will show up as a homozygous change. Taking everything together, the way toward getting a succession from whole blood and the examination may require 4–5 days, with certification using another PCR based test before the change is accounted for. The only drawbacks of using sequencing as a routine investigation technique are the cost and time taking examination compared to PCR. Sequencing is a multistage procedure requiring PCR intensification, cycle sequencing and precipitation before the sequence can be distinguished. After this the sequence ought to be researched and checked and any movements noted. Notwithstanding the grouping examination programming is available it is not 100% compelling at identifying heterozygotes.

Next-generation sequencing (NGS) allow the generation of immense measures of genomic data to uncover the genetic constitution of people and to evaluate potential health risks. NGS has been commonly utilized for non-invasive prenatal diagnosis and novel mutation detection in thalassemia [57, 58].

Microarray analysis procedures are utilized in deciphering the information produced from probes DNA (Gene chip examination), RNA, and protein microarrays, which permit scientist to explore the expression of large number of gene in many cases, an organism's whole genome in a single step experiments. The complementary sequences will bind to each other. The unknown DNA's are cut into pieces by restriction endonucleases and these DNA pieces are label with fluorescent markers. These are then allowed to react with tests of the DNA chip [59–61].

Microarray analysis of gene expression has formed into a great tool for the characterization of various pathophysiological processes. The fundamental idea is that RNA isolated from tissue is hybridized to probes for specific genes that are fixed in a grid in small microscopic spots. The microarray is a quick, simple to perform, and precise strategy for concurrent identification of α and β-thalassemias. But, this technique needs should be improved and approved in a bigger number of specimens with hemoglobinopathies before further routine laboratory use [62–64].

3.4 Prenatal screening

The prenatal diagnosis based on DNA by amniocentesis and CVS sampling is required to detect the genetic abnormalities of the foetus. Now a day's NIPT is a technique to identify the genetic abnormalities of the foetus. This testing examines the small remains of foetus DNA that are circulating in a pregnant mother's blood. An appropriate lab conclusion is urgent for describing the various types of thalassemia with a significant association for prevention and treatment. Because of population migration and mixing of the gene pool of different populations in many immigration countries as well as regions the hemoglobiopathies and thalassemia are more prevalent over their [65–68].

The importance of prenatal diagnosis comes in the diagnostic field as it helps and early diagnoses the growing foetus in the mother wombs for thalassemia and hemoglobinopathies. It plays impartment role in the eradication of these genetic disorders such as β-thalassemia major, sickle cell disease and hemoglobin Bart's nonimmunehydropsfetalis [69, 70]. The prenatal diagnosis includes the investigation of fetal material from chorionic villi, amniotic liquid, string blood, and fetal DNA in maternal dissemination. In spite of the way that examination of fetal hemoglobin types is successfully performed by means of robotized HPLC, it is assessable through assessment of fetal blood got by cordocentesis and the technique is inclined to mistake because of mixing of sample by maternal tissue [71].

Advances in molecular testing have worked with the assurance of complex thalassemias and hemoglobinopathies saw in ethnically varying population. Comprehensive screening programs highlighted recognizing carriers and offering prenatal diagnosis in pregnancies for thalassemia have been incorporated in Canada and European countries [72, 73]. Different procedures have been implimented to distinguish thalassemia like genotyping assay, genotyping measur next-generation sequencing and mass spectrometry [74–76]. The methods are as yet testing; consequently, more investigations are expected to create and approve them and eventually lead to proficient, exact and concrete non-invasive prenatal diagnosis of thalassemia and hemoglobinopathies [77].

4. Cost comparison for testing

	Low cost techniques	High cost techniques
Hematological and Biochemical Techniques	MCV and MCH	Mass spectrometry
	CBC	
	RDW	
	Electrophoresis	
	HPLC	
Molecular Techniques	ASO, RDB,	MLPA
	ARMS PCR	Direct sequence
	RE PCR/RFLP	Microarray
	Gap-PCR	HRM
	DGGE and SSCP	

5. Point of care (POC) testing for thalassemia

As there is a saying that prevention is better than cure. The thalassemia and hemoglobinopathies are genetic disorder and due to the migration of population in the different regions and endogamy the new combinations of thalassemia hemoglobinopaty are arising fast. For the eradication and further treatment and management of thalassemia disease an effective diagnostic test at the point of care (POC) is the need of hour.

As thalassemia has been spread worldwide an early diagnosis, economical test, awareness programmes and prenatal screening will be a milestone for the eradication of this genetic disorder and to reduce burden of the health sector of a country subsequently the economics.

The initial hematological, biochemical screening to the advance molecular testing under one roof will not only help to diagnose the thalassemia patients but shall be also helpful in the treatment and management of the disease (**Figure 1**). The objective of POC for testing thalassemia will be achieved if these quick testing methods like NESTROFT. This has to be in the approachable distance to the patients.

The government should include the thalassemia screening as a mandatory tool for screening. The best example Italy and Cyprus there with the initiation of prenatal diagnosis and screening of the carrier now this is a thalassemia free country.

Likewise in India, the Uttar Pradesh state government has initiated the task by giving free of cost of screening of thalassemia carrier, blood transfusion and iron chelators to several medical colleges. The Indian government has provided the HPLC machines to provide the screening of the carrier of thalassemia hemoglobinopathies. Basically the eradication, treatment and management of thalassemia are joint efforts of government, stake holder, policy makers, pediatrician, pathologist, transfusion medicine and geneticist.

Author details

Nitu Nigam[1]*, Prithvi Kumar Singh[1], Suhasini Bhatnagar[2], Sanjay Kumar Nigam[3] and Anil Kumar Tripathi[4]

1 Department of Centre for Advance Research (Cytogenetics Lab), King George's Medical University, Lucknow, Uttar Pradesh, India

2 RUS Industries, Ghaziabad, Uttar Pradesh, India

3 Department of Pathology, Saraswati Medical College, Unnao, Uttar Pradesh, India

4 Department of Clinical Hematology, King George's Medical University, Lucknow, Uttar Pradesh, India

*Address all correspondence to: nigamnitu@gmail.com

References

[1] Old J. Screening and genetic diagnosis of hemoglobin disorders. Blood Reviews. 2003:43.

[2] Old JM. Screening and genetic diagnosis of haemoglobinopathies. Scand J Clin Lab Invest. 2006;66:1-16.

[3] Old J, Harteveld CL, Traeger-Synodinos J, et al. Prevention of Thalassaemias and Other Haemoglobin Disorders: Volume 2: Laboratory Protocols [Internet]. 2nd edition. Nicosia (Cyprus): Thalassaemia International Federation; 2012. Chapter 5, MOLECULAR DIAGNOSIS. Available from: https://www.ncbi.nlm.nih.gov/books/NBK190571

[4] Brown TA. Genomes. 2nd edition. Oxford: Wiley-Liss; 2002. Chapter 7, Understanding a Genome Sequence. Available from: https://www.ncbi.nlm.nih.gov/books/NBK21136/

[5] Colah RB, Gorakshakar AC. Lokeshwar MR, Shah NK, Agarwal B, Suchdeva A, editors. Structural hemoglobinopathies. *IAP speciality series on Pediatric Hematolgy and Oncolgy of Indian Academy of Pediatrics.* 2006:151–161.

[6] Ryan K, Bain BJ, Worthington D, James J, Plews D, Mason A, et al. Significant haemoglobinopathies: Guidelines for screening and diagnosis. Br J Haematol. 2010;149:35-49.

[7] Nigam N, Kushwaha R, Yadav G, Singh PK, Gupta N, Singh B, Agrawal M, Chand P, Saxena SK, Bhatt MB. A demographic prevalence of β Thalassemia carrier and other hemoglobinopathies in adolescent of Tharu population. J Family Med Prim Care 2020;9:4305-4310.

[8] Ghosh K, Colah R, Manglani M, et al. Guidelines for screening, diagnosis and management of hemoglobinopathies. Indian J Hum Genet. 2014;20(2): 101-119.

[9] Bain BJ. 2nd ed. Oxford: Wiley Blackwell; 2006. Hemoglobinopathy diagnosis. ISBN: 979 - 1 - 4051-3516-0.

[10] Lewis S, Bain B, Bates I. 10th ed. UK: Churchill Livingstone; 2006. Dacie and Lewis practical hematology.

[11] Mohanty D, Colah R, editors. 1st ed. Mumbai: Bhalani Publishing House; 2008. Laboratory Manual for screening diagnosis and molecular analysis of hemoglobinopathies and red cell enzymopathies.

[12] Clark BE, Thein SL. Molecular diagnosis of haemoglobin disorders. Clin Lab Haematol. 2004;26(3):159-176.

[13] Lee YK, Kim HJ, Lee K, Park SH, Song SH, Seong MW, Kim M, Han JY. Recent progress in laboratory diagnosis of thalassemia and hemoglobinopathy: a study by the Korean Red Blood Cell Disorder Working Party of the Korean Society of Hematology. Blood Res. 2019;54(1):17-22.

[14] Clarke GM, Higgins TN. Laboratory investigation of hemoglobinopathies and thalassemias: review and update. Clin Chem. 2000; 46:1284-1290.

[15] Recommendations of a system for identifying abnormal hemoglobins. By the International Committee for Standardization in Hematology. Blood. 1978; 52:1065-1067.

[16] Ryan K, Bain BJ, Worthington D, et al. Significant haemoglobinopathies: guidelines for screening and diagnosis. Br J Haematol. 2010; 149:35-49.

[17] Kim SY, Lee SH, Cho SI, et al. Molecular identification of the novel Gγ-β hybrid hemoglobin: Hb Gγ-β Ulsan (Gγ through 13; β from 19). Blood Cells Mol Dis. 2010; 45:276-279.

[18] Mahdieh N, Rabbani B. An overview of mutation detection methods in genetic disorders. Iran J Pediatr. 2013;23(4):375-388.

[19] Cao, A., Galanello, R. Beta-thalassemia. Genet Med 12, 61-76 (2010).

[20] Thein SL. The molecular basis of β-thalassemia. Cold Spring Harb Perspect Med. 2013;3(5):a011700. Published 2013 May 1. doi:10.1101/cshperspect.a011700.

[21] Hardison RC. Evolution of hemoglobin and its genes. Cold Spring Harb Perspect Med. 2012;2(12):a011627. Published 2012 Dec 1. doi:10.1101/cshperspect.a011627

[22] Huisman T, Carver M, Baysal E, Efremov G. A Database of Human Hemoglobin Variants and Thalassemias. State College, PA: The Pennsylvania State University;2013. Accesed December, 8, 2018. at http://globin.csc.psu.edu/cgi-bin/hbvar/query_vars3.

[23] Conner B.J., Reyes A.A., Morin C., Itakura K., Teplitz R.L. & Wallace R.B. (1983) Detection of sickle cell bs -globin allele by hybridization with synthetic oligonucleotides. Proceedings of the National Academy of Sciences of the United States of America 80, 278-282.

[24] Wallace R.B., Johnson M.J., Hirose T., Miyake T., Kawashima E.H. & Itakura K. (1981) The use of synthetic oligonucleotides as hybridization probes. II: Hybridisation of oligonucleotides of mixed sequence to rabbit b-globin DNA. Nucleic Acids Research 9, 879 894.

[25] Orkin. S.H.. Sexton, J.P., Cheng. T.C., Goff, S.C.. Giardina, P.V.J. & Kazazian. H.H., Jr (1983) TATA box transcription mutation in βthalassemia. Nucleic Acids Research. 11, 4721-4734.

[26] Pirastu M, Kan YW, Cao A, Conner BJ, Teplitz RL, Wallace RB. Prenatal diagnosis of beta-thalassemia. Detection of a single nucleotide mutation in DNA. N Engl J Med. 1983 Aug 4;309(5):284-287.

[27] Ristaldi M.S., Pirastu M., Rosatelli C., Cao A. Prenatal diagnosis of β-thalassaemia in Mediterranean populations by dot blot analysis with DNA amplification and allele specific oligonucleotide probes. Prenatal Diagnosis. 1989;9:629-638.

[28] Saiki R.K., Walsh P.S., Levenson C.H., Erlich H.A. Genetic analysis of amplified DNA with immobilized sequence-specific oligonucleotide probes. Proceedings of the National Academy of Sciences, USA. 1989;86:6230-6234.

[29] Newton C.R., Graham A., Heptinstall L.E. Analysis of any point mutation in DNA. The amplification refractory mutation system (ARMS). Nucleic Acids Research. 1989;17:2503-2516.

[30] Old J.M., Varawalla N.Y, Weatherall D.J. The rapid detection and prenatal diagnosis of β thalassaemia in the Asian Indian and Cypriot populations in the UK. Lancet. 1990;336:834-837.

[31] Old JM, Khan SN, Verma, et al. A multi-centre study to further define the molecular basis of beta-thalassemia in Thailand, Pakistan, Sri Lanka, Mauritius, Syria, and India, and to develop a simple molecular diagnostic strategy by amplification refractory mutation system polymerase chain reaction. Hemoglobin. 2001;25:397.

[32] Saxena R., Jain P.K., Thomas E, Verma I.C. Prenatal diagnosis of β-thalassaemia: experience in a developing country. Prenatal Diagnosis. 1998;18:1-7.

[33] Baig SM. Molecular diagnosis of beta-thalassemia by multiplex

ARMS-PCR: a cost effective method for developing countries like Pakistan. Prenat Diagn. 2007;27:580-581.

[34] Tan J.A., Tay J.S, Lin L.I., et al. The amplification refractory mutation system (ARMS): a rapid and direct prenatal diagnostic technique for β-thalassaemia in Singapore. Prenatal Diagnosis. 1994;14:1077-1082.

[35] Ye S, Dhillon S, Ke X, Collins AR, Day IN. An efficient procedure for genotyping single nucleotide polymorphisms. Nucleic Acids Res. 2001;29(17):E88-e88.

[36] Tan JA, Tay JS, Lin LI, Kham SK, Chia JN, Chin TM, Aziz NB, Wong HB. The amplification refractory mutation system (ARMS): a rapid and direct prenatal diagnostic technique for beta-thalassaemia in Singapore. Prenat Diagn. 1994;14(11):1077-1082.

[37] Linderman R., Hu S.P., Volpato F, Trent R.J. Polymerase chain reaction (PCR) mutagenesis enabling rapid nonradioactive detection of common β-thalassaemia mutations in Mediterraneans. British Journal of Haematology. 1991;1991;78:100-104.

[38] Varawalla NY, Fitches AC, Old JM. Analysis of beta-globin gene haplotypes in Asian Indians: origin and spread of beta-thalassaemia on the Indian subcontinent. Hum Genet. 1992;90:443-449.

[39] Weatherall DJ, Clegg JB. Inherited haemoglobin disorders: an increasing global health problem. Bull World Health Organ. 2001;79(8):704-712.

[40] Faa V., Rosatelli M. C., Sardu R., Meloni A., Toffoli C., Cao A. A simple electrophoretic procedure for fetal diagnosis of b-thalassaemia due to short deletions. Prenat Diag. 1992;12:903-908.

[41] Chong S. S., Boehm C. D., Higgs D. R., Cutting G. R. Single-tube multiplex-PCR screen for common deletional determinants of α-thalassemia. Blood. 2000;95:360-362.

[42] Liu Y.T., Old J.M., Miles K., Fisher C.A., Weatherall D.J. & Clegg J.B. Rapid detection of alpha-thalassaemia deletions and alpha-globin gene triplication by multiplex polymerase chain reactions. British Journal of Haematology 2000; 108, 295-299.

[43] Losekoot M, Fodde R, Harteveld CL, van Heeren H, Giordano PC, Bernini LF. Denaturing gradient gel electrophoresis and direct sequencing of PCR amplified genomic DNA: a rapid and reliable diagnostic approach to beta thalassaemia. Br J Haematol. 1990 Oct;76(2):269-274.

[44] Losekoot, M., Fodde, R., Harteveld, C. L., van Heeren, H., Giordano, P. C., Went, L. N., and Bernini, L. F. (1991). Homozygous β+−thalassaemia owing to a mutation in the cleavage-polyadenylation sequence of the human β-globin gene. J. Med. Genet. 28:252-255.

[45] Savage, D.A., Wood, N.A., Bidwell, J.L., Fitches, A., Old, J.M. & Hui, K.M. (1995) Detection of b-thalassaemia mutations using DNA heteroduplex generator molecules. British Journal of Haematology, 90, 564-571.

[46] Pornprasert S, Phusua A, Suanta S, et al. Detection of alpha-thalasemia-1 Southeast Asian type using real-time gap-PCR with STBR green1 and high resoulution melting analysis. Eur J Haematolo. 2008;80(6):510-514.

[47] Prajantasen T, Fucharoen S, Fucharoen G. High resolution melting analytical platform for rapid prenatal and postnatal diagnosis of β-thalassemia common among Southeast Asian population. Clin Chim Acta. 2015;441:56-62.

[48] Harteveld Cl, Voskamp A, Phylipsen M, et al. Nine unknown

rearrangements in 16p13.3 and 11p15.4 causing alpha- and beta-thalassaemia characterised by high resolution multiplex ligation-dependent probe amplification. J Med Genet. 2005;42:922-931.

[49] Liu JZ, Han H, Schoulten JP, et al. Detection of alpha-thalassemia in China by using multiplex ligation-dependent probe amplification. Hemoglobin. 2008;32(6):561-571.

[50] Harteveld CL, Refaldi C, Cassinerio E, et al. Segmental duplications involving the alpha-globin gene cluster are causing beta-thalassemia intermedia phenotypes in beta-thalassemia heterozygous patients. Blood Cells Mol Dis. 2008;40(3):312-316.

[51] Koonin EV, Galperin MY. Sequence - Evolution - Function: Computational Approaches in Comparative Genomics. Boston: Kluwer Academic; 2003. Chapter Principles and Methods of Sequence Analysis. Available from: https://www.ncbi.nlm.nih.gov/books/NBK20261/

[52] Fakher R, Bijan K, Taghi AM. Application of diagnostic methods and molecular diagnosis of hemoglobin disorders in Khuzestan province of Iran. Indian J Hum Genet. 2007;13(1):5-15. doi:10.4103/0971-6866.32028

[53] Bajwa H, Basit H. Thalassemia. [Updated 2021 Jan 24]. In: StatPearls [Internet]. Treasure Island (FL): StatPearls Publishing; 2021 Jan-. Available from: https://www.ncbi.nlm.nih.gov/books/NBK545151/.\

[54] Sanger F, Donelson JE, Coulson AR, Kossel H, Fischer D. Use of DNA polymerase I primed by a synthetic oligonucleotide to determine a nucleotide sequence in phage fl DNA. Proc Natl Acad Sci U S A. 1973;70(4):1209-1213.

[55] Sanger F, Nicklen S, Coulson AR. DNA sequencing with chain-terminating inhibitors. Proc Natl Acad Sci U S A. 1977;74(12):5463-5467.

[56] Bwayo D, Kaddumukasa M, Ddungu H, Kironde F. Prevalence of glucose-6-phosphate dehydrogenase deficiency and its association with Plasmodium falciparum infection among children in Iganga distric in Uganda. BMC Res Notes. 2014 Jun 18;7:372.

[57] Papasavva T, van Ijcken WF, Kockx CE, et al. Next generation sequencing of SNPs for non-invasive prenatal diagnosis: challenges and feasibility as illustrated by an application to β-thalassaemia. Eur J Hum Genet 2013;21:1403-1410.

[58] Xiong L, Barrett AN, Hua R, et al. Non-invasive prenatal diagnostic testing for β-thalassaemia using cell-free fetal DNA and next generation sequencing. Prenat Diagn 2015;35:258-265.

[59] Zesong L, Ruijun G, Wen Z. Rapid detection of deletional alpha-thalassemia by an oligonucleotide microarray. American journal of Hematology. 2005;80:306-308.

[60] Bang-Ce Y, Hongqiong L, Zhuanfong Z, et al. Simultaneous detection of alpha-thalassemia and beta-thalassemia by oligonucleotide microarray. Haematologica. 2004;89:1010-1012.

[61] Gemignani F, Perra C, Landi S, et al. Reliable detection of beta-thalassemia and G6PD mutations by a DNA microarray. Clinical Chemistry. 2002;48:2051-2054.

[62] Van Moorsel CH, van Wijngaraarden EE, Foldkema IF, et al. beta-Globin mutation detection by tagged single-base extension and hybridization to universal glass and flow-through microarrays. European Journal Human Genetics. 2004;12:567-573.

[63] Lu Y, Kham SK, Tan PL, et al. Arrayed primer extension: a robust and reliable genotyping platform for the

diagnosis of single gene disorders: beta-thalassemia and thiopurine methyltransferase deficiency. Genet Test. 2005;9:212-219.

[64] Cremonesi L, Ferrari M, Giordano PC, et al. An overview of current microarray-based human globin gene mutation detection methods. Hemoglobin. 2007;31(3):289-311.

[65] Sayani FA, Kwiatkowski JL. Increasing prevalence of thalassemia in America: implications for primary care. Ann Med. 2015;47:592-604.

[66] Anwar WA, Khyatti M, Hemminki K. Consanguinity and genetic diseases in North Africa and immigrants to Europe. Eur J Public Health. 2014;24(Suppl 1):57-63.

[67] Jahng J, Yoon KH. A family with a hemoglobin E variant including a Thai immigrant woman in Korea. Ann Lab Med. 2017;37:71-73.

[68] Lee HS, Lee DY, Kim HJ, Lee IS. Two cases of alpha-thalassemia in Korean children from multicultural family. Clin Pediatr Hematol Oncol. 2011;18:136-139.

[69] Li DZ, Yang YD. Invasive prenatal diagnosis of fetal thalassemia. Best Pract Res Clin Obstet Gynaecol. 2017;39:41-52.

[70] 27.Seale TW, Rennert OM. Prenatal diagnosis of thalassemias and hemoglobinopathies. Ann Clin Lab Sci. 1980;10:383-394.

[71] Sanguansermsri T, Thanaratanakorn P, Steger HF, et al. Prenatal diagnosis of hemoglobin Bart's hydrops fetalis by HPLC analysis of hemoglobin in fetal blood samples. Southeast Asian J Trop Med Public Health. 2001;32:180-185.

[72] Cousens NE, Gaff CL, Metcalfe SA, Delatycki MB. Carrier screening for beta-thalassaemia: a review of international practice. Eur J Hum Genet. 2010;18:1077-1083.

[73] Langlois S, Ford JC, Chitayat D CCMG Prenatal Diagnosis Committee; SOGC Genetics Committee. Carrier screening for thalassemia and hemoglobinopathies in Canada. J Obstet Gynaecol Can. 2008;30:950-959.

[74] Papasavva T, van Ijcken WF, Kockx CE, et al. Next generation sequencing of SNPs for non-invasive prenatal diagnosis: challenges and feasibility as illustrated by an application to β-thalassaemia. Eur J Hum Genet. 2013;21:1403-1410.

[75] Li Y, Di Naro E, Vitucci A, et al. Size fractionation of cell-free DNA in maternal plasma improves the detection of a paternally inherited beta-thalassemia point mutation by MALDI-TOF mass spectrometry. Fetal Diagn Ther. 2009;25:246-249.

[76] Breveglieri G, Travan A, D'Aversa E, et al. Postnatal and non-invasive prenatal detection of β-thalassemia mutations based on Taqman genotyping assays. PLoS One. 2017;12:e0172756.

[77] Zafari M, Kosaryan M, Gill P, et al. Non-invasive prenatal diagnosis of β-thalassemia by detection of the cell-free fetal DNA in maternal circulation: a systematic review and meta-analysis. Ann Hematol. 2016;95:1341-1350.

Chapter 8

Pulmonary Hypertension in Thalassemia Patients

Ahmed Shemran Mutlaq Alwataify, Sabih Salih Alfatlawy and Yahia Abid Alshahid Altufaily

Abstract

Pulmonary hypertension (PH) is defined in children as a mean pulmonary arterial pressure (PAP) greater than 25 mmHg at rest or 30 mmHg during physical activity, with increased pulmonary artery capillary wedge pressure and an increased pulmonary vascular resistance greater than 3 Wood units × M^2. it is the main cause of morbidity and mortality in the group of thalassemia, if no treatment leads to right ventricular heart failure and death. The development of pulmonary arterial hypertension (PAH) is assumed to be the result of many multifactorial pathogenic mechanisms including chronic hemolysis, iron overload, hypercoagulability, and erythrocyte dysfunction as a result of splenectomy, inflammation and nitric oxide (NO) depletion. PAH symptoms are non-specific, their signs consist of right ventricular lift, an accentuated pulmonary component of the second heart sound, a (gallop rhythm) right ventricular third heart sound, and parasternal heave meaning a hypertrophied right ventricle. The diagnosis of PAH requires a clinical suspicion based on symptoms and physical examination. Echocardiography is frequently used to screen for PAH, monitor progression over time and allow identification of patients for whom diagnostic right heart catheterization (RHC) is warranted and its treatment includes hemoglobinopathy specific treatment and PAH specific therapy.

Keywords: thalassemia, pulmonary hypertension, splenectomy, iron overload, blood transfusion, hydroxyurea, hypercoagultion, specific therapy

1. Introduction

Thalassemia are groups of autosomal recessive inherited disorders resulting from reduced or defects in one or more of the hemoglobin (Hb) chains synthesis [1]. The thalassemia syndrome is classified to which of the globin chains α or β is affected [2]. The α- and β-thalassemia are the main types of thalassemia and sickle cell thalassemia [3].

Thalassemia are found with the increased prevalence in the Mediterranean and Middle East, is caused by mutations on the chromosomes 11 and 16 for the cases of β- and α-thalassemia respectively with more than 150 different mutations [4]. β-Thalassemia presents in three clinical phenotypes, called thalassemia major, minor, and intermedia. β-Thalassemia major is resulting from homozygous or compound heterozygous to β-thalassemia and usually has early presentation and it depends on blood transfusion [1], reduced survival, has multi organs complications, frequent hospitalization and need lifelong management [5]. It is the most

IntechOpen

severe form in patients with defective in two β-globulin genes and severe reduction in production of β-globulin [3]. β-Thalassemia minor, on the other hand, is an asymptomatic condition due to heterozygous to a β-thalassemia defect [1]. β-Thalassemia intermedia includes patients with a wide series of phenotypes [6] which is more severe than thalassemia minor and milder than β-thalassemia major [7]. β-Thalassemia intermedia patients with two defective genes is characterized by mild to moderate reduction in β-globulin production [4]. Thalassemia intermedia associated with lack of regular blood transfusion and iron chelation therapy leading to serious specific complications like gall and renal stone, leg ulcer and increased thrombophilia and right heart failure [8].

Its pathology is characterized by reduced synthesis of hemoglobin (Hb) and the survival of red blood cell (RBC), caused by from excessing the unaffected globin chain [9]. This abnormal alpha- to beta-chain ratio lead to precipitation of the unpaired chains which result in destruction of red blood cell precursors in the bone marrow called ineffective erythropoiesis and in circulation named hemolysis [10, 11]. Ineffective erythropoiesis causes expanded of marrow cavities leading to distortion of the cranium, facial and long bones and also produced lymphadenopathy, hepatosplenomegaly because of activity in extramedullary hematopoietic sites [12]. Children with β-thalassemia appear well at birth, then developed anemia that worse with time, if left no treatment resulting in early death [13] due to high output heart failure [14].

Therapy involves early and regular transfusion to maintain hemoglobin levels of at least 9–10 g/dl to make for improving growth and development and reduced hepatosplenomegaly in addition to bone deformities [15], regular blood transfusion therapy lead to prolonged survival [16], decreased the severity of the disease [17] and hemolysis because of chronic transfusion might ameliorate the ineffective erythropoiesis, and in that way reverse the pulmonary vasoconstriction and pulmonary hypertension (PH) [18].

So, regular blood transfusion, iron chelation drugs and hydroxyurea therapy are frequently working resulted in expressively improved survival, and may extend their life to the adulthood [19].

2. Mechanism of iron toxicity

It results from prolonged iron absorption especially thalassemia intermedia or repeated blood transfusion in thalassemia major. Iron is highly reactive and easily interchanges between two states iron III and iron II in a method which results in the loss as well as gain of electrons leading to generation of unsafe free radicals which damaged lipid membranes, organelles and deoxyribonucleic acid (DNA) leading to cell death and fibrosis is developed [20]. In healthy, iron is binding to transferrin so kept safe, but in iron overload, the transferrin capacity to bind iron is exceeded within cells and plasma resulting in free iron leading to damage of many tissues which fatal unless iron chelation treatment [9]. Even though, blood transfusion improved the severity of the disease, it resulted in a positive iron balance and secondary iron overload, leading to damage and dysfunction of vital organs in the second decade of life [17] (each bottle of blood transfusion contains 200 mg of iron [14]).

3. Pulmonary arterial hypertension

Pulmonary arterial hypertension (PAH) is defined generally as high blood pressure in the heart to lung system that delivered fresh oxygenated blood to the heart while returning used (oxygen depleted) blood back to the lung [21]. It is defined in children as a mean pulmonary arterial pressure (PAP) greater than 25 mmHg at rest or 30 mmHg during physical activity with increased pulmonary artery capillary wedge pressure likely (due to left ventricle diastolic dysfunction as resulting from chronic iron overload) and an increased pulmonary vascular resistance (from thrombotic pulmonary arteriopathy) greater than 3 Wood units × M^2 [22].

PAH is a progressive disease [23] associated with hemoglobinopathies. It is a not uncommon in thalassemia patients, caused by pulmonary hemosiderosis [8]. It is the important cause of morbidity and mortality in this group of the patients [24].

PAH has been reported as one of the common cardiac complications in β-thalassemia patients [25]. It is characterized by vasoconstriction, and progressive increases in the mean pulmonary artery pressure and pulmonary vascular resistance, if no treatment leads to right ventricular heart failure and death [26].

A universal classification system has been used to describe the various types of pulmonary hypertension [27]:

1. Pulmonary arterial hypertension called (idiopathic): is inherited; is resulted from drugs or toxins; connective tissue disease, human immunodeficiency virus (HIV) infection, liver disease, congenital heart disease or is caused by conditions that affect the pulmonary small blood vessels [21].

2. Pulmonary hypertension due to left heart disease.

3. Pulmonary hypertension due to lung diseases and/or hypoxia.

4. Chronic thromboembolic pulmonary hypertension.

5. Pulmonary hypertension with indistinct multifactorial mechanisms including chronic hemolytic anemia shares many of the included features of group 1 and 5 due to its multifactorial mechanisms [19, 28].

4. Pathophysiology

The PAH in thalassemia syndromes is assumed to result from many multifactorial mechanisms, it includes chronic hemolysis, iron overload, hypercoagulability [29] and erythrocyte dysfunction resulting from splenectomy [19], inflammation and nitric oxide (NO) depletion free Hb inactivate nitric oxide (NO) and vasodilatory properties in pulmonary circulation and the hemolysis resulting in release of arginase enzyme which predominantly drive L-arginine to ornithine rather than NO production, therefore higher arginase activity and lower arginine bioavailability founded in thalassemia as L-arginine provides the substrate of both NO synthesis and arginase [19] as well as enhance activation of platelets and increased endothelin 1 release causing vasculopathy which are main pathogenetic mechanisms [30].

Pulmonary arterial hypertension is characterized histopathologically by vasoconstriction, vascular proliferation, intravascular thrombosis, and remodeling of the vessels walls [26].

Nitric oxide caused vasodilation and increased blood flow through inhibiting expression of endothelial adhesion molecule, proliferation of vascular smooth muscle, platelet aggregation and blood coagulation [31]. As intravascular hemolysis, arginase concurrently released from red cells into blood plasma, converts plasma L-arginine to ornithine result in high level of ornithine which produced smooth muscle proliferation and collagen synthesis resulting in vascular remodeling and thickening in intimal leading to vascular constriction, endothelium abnormality, and intravascular thrombosis [32].

Splenectomy has been an important risk factor in the development of PAH [11] as spleen is considered as filter for erythrocytes that are damaged and other circulating blood cells. so splenectomy is caused in activation of platelets, abnormal in erythrocyte aggregation, thrombin generation acceleration and released procoagulant factors, possibly it has a larger number of abnormal erythrocytes or breakdown products of erythrocyte in the circulation and associated with excess in release of early nucleated erythrocytes leading more expression of adhesion molecules which promote local clot formation particularly in the presence of systemic hypercoagulability [33]. In addition, there was shorter life span of platelets in both splenectomized and nonsplenectomized patients with thalassemia than in non thalassemia splenectomized patients [33]. These factors may affect the ability of intravascular thrombosis producing changes in the pulmonary vasculature leading to PAH [19].

A hypercoagulable state has another finding among hemoglobinopathies. The incidence of clinically evident thromboembolism in β-thalassemia is 1–4%, with a majority of events occurring prior 30 years of age [19]. It results from a range of abnormalities, including erythrocyte aggregation, the performed splenectomy, platelets activation, abnormal circulating factors, some coexistent genetic coagulation defects, endothelial dysfunction and vasculopathy (**Figure 1**) [34]. The multifactorial mechanisms described earlier can work both independently and in combination with other mechanisms or organ dysfunction to lead clinically apparent PAH [19].

PAH is more common in thalassemia intermedia than in thalassemia major, and it may cause heart complications in patients with aging of more than 30 years [35]. Its iron load occurs as increased uptake of gastrointestinal absorption of iron due to hepcidin suppression [14] making more susceptible to pulmonary arterial hypertension [35].

In β-thalassemia major, pulmonary hypertension correlates with the severity of hemolysis [19], although transfusion therapy improved the disease severity, it resulted in a positive iron balance and secondary haemosiderosis, often leading to vital organ damage and dysfunction in the second decade of life [36], while splenectomy has important role in both types [19], yet in thalassemia patients whose disease is well treated by chronic transfusion therapy, the development of pulmonary hypertension is linked to heart dysfunction and the subsequent toxic effects of iron overload instead of hemolysis [19].

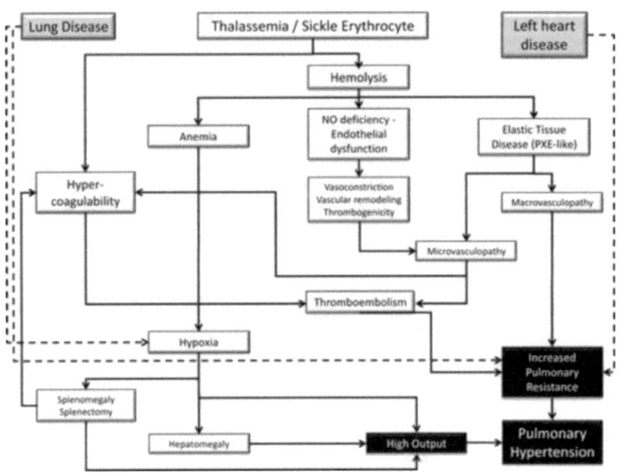

Figure 1.
The pathophysiology of pulmonary hypertension in thalassemia [34].

5. Risk factors

The risk of development of PHT can increase older age group (more often in people older than 30 years) but idiopathic PAH is more common in younger adults. Chronic iron overload and increased pulmonary vascular resistance from thrombotic pulmonary arteriopathy [29], splenectomy, hepatitis C and previous venous thromboembolism, markedly increased levels of peripheral nucleated red blood cells, platelet counts and serum ferritin levels [25] and splenectomy [37].

6. Epidemiology

The epidemiology of idiopathic PAH is about 125–150 deaths per year in the United States, and worldwide the incidence is similar to the united states at 4 cases per million. However, in the France, the incidence were 6 cases per million. Females have a higher incidence rate than males [38], thrombi in small pulmonary arteries reached in 44% of splenectomized Hb E/B thalassemia. In Greece 10% occurred in in thalassemia major and more than 50% in thalassemia intermedia, in Ardabil, Iran 41.4% in intermedia and 14.7% in thalassemia major [8], while its prevalence in Thailand was 43% [39] and Babylon, Iraq reached to 15.5% [40].

PHT increased its incidence about 2.45 times chance in patients with serum ferritin of more than 1000 ng/ml [39].

Pulmonary vascular changes with micro-thrombo-embolism founded in splenectomized thalassemia 54% in Germany [41], 33.3% in Babylon, Iraq [40] and only 16% and 13.4% in non splenectomized patients [40, 41] respectively.

PAH has a frequent finding in patients with hemoglobinopathies, on the other hand, The reported prevalence differs in the different conditions state and according to the screening method used. In thalassemia intermedia, PAH has been known as the most cardiovascular finding and the commonest cause of heart failure. In homozygous sickle cell anemia, PAH is a detected frequently and it has been thought a major factor determining the prognosis [42]. In sickle-thalassemia, a heterozygous state with 1 thalassemia and 1 sickle allele, PAH has been detected with a relatively lower frequency and absolutely lesser severity [34].

7. Signs and symptoms

PAH symptoms are non-specific: either early symptoms which include shortness of breath or exertional dyspnea, fatigue, weakness, chest pain, upper abdominal pain and decreased appetite, while later symptoms include light-headedness (syncope) and less frequently cough, fainting, edema of ankle or leg, Rarely hemoptysis, hoarseness, arrhythmias and cyanosis [43].

Physical examination: the signs consist of right ventricular lift, an accentuated pulmonary component of the second heart sound, a (gallop rhythm) right ventricular third heart sound, and parasternal heave meaning a hypertrophied right ventricle. Right sided heart failure causes signs of systemic congestion include jugular venous distension, ascites, and hepatojugular reflux, hepatomegaly and or splenomegaly, the evidence of tricuspid insufficiency and pulmonic regurgitation [44].

World Health Organization (WHO) functional classification in pulmonary hypertension (stages of PH) [45]:

I. without limitation of physical activity (ordinary physical activity does not produce undue dyspnea or fatigue, chest pain or near syncope).

II. slight limitation of physical activity (comfortable at rest; its ordinary physical activity causes undue dyspnea or fatigue, chest pain or near syncope).

III. marked limitation (comfortable at rest; less than ordinary physical activity causes undue dyspnea or fatigue, chest pain or near syncope).

IV. inability to carry out any physical activity without symptoms (the patients presented with signs of right sided heart failure, dyspnea and /or fatigue which manifested at rest and discomfort is increased by any physical activity).

Cardiac evaluation through chest radiograph, electrocardiogram (ECG) and echocardiography. Chest radiography (CXR) was performed in erect position, in the posterior-anterior (PA) view to look for cardiac size, pulmonary vascularity and pulmonary conus [45].

8. Complications

It includes:

1. Right-sided heart dilatation and heart failure: PAH is a deadly disease in which vasoconstriction and vascular remodeling both lead to a progressive increase in pulmonary vascular resistance. Although the increment in afterload is the first trigger for RV adaptation. Neurohormonal signaling, oxidative stress, inflammation, ischemia, and cell death may contribute to the development of RV dilatation and failure [46].

2. Blood clots: Intravascular thrombosis are found in the small distal pulmonary arteries. Dysregulation of coagulation, platelet function, and endothelial cells may contribute to a prothrombotic state [47].

3. Arrhythmia: Pulmonary hypertension can cause arrhythmias, atrial flutter and fibrillation [48].

9. Diagnosis

The evaluation of pulmonary hypertension in children required a comprehensive work up to confirm the diagnosis, assess disease severity and rule out secondary cause, so appropriate treatment course can be initiated [45].

The diagnosis requires a clinical suspicion based on symptoms and physical examination and review of a comprehensive set of investigations to confirm haemodynamic criteria are met and to describe the etiology, the functional and haemodynamic severity of the condition [49].

9.1 Transthoracic Doppler echocardiography (TTE)

It is non-invasive, an excellent screening tool with suspected PAH, it is widely available and relatively inexpensive, it is frequently used to screen for PAH and monitor progression over time and allow identification of patients for whom diagnostic RHC is warranted [50]. It documents cardiac anatomy, right ventricular size and function, left ventricular diastolic and systolic function, morphology and function of valves, and the presence of pericardial effusion or a patent foramen ovale [22].

Transthoracic Doppler echocardiography (TTE) can assess pulmonary artery systolic pressure (PASP) and give additional information about the cause and consequences of PAH [51]. It also can provide functional and structural assessment of the heart and estimate of pulmonary hemodynamics, it is widely available and tolerated by children [52]. Echocardiographic assessment is essential for the effective modality of PAH management patients as part of a good approach to therapy and prognosis (presence of pericardial effusion and increasing right ventricle to left ventricle systole ratio is associated with an increasing hazard for a clinical event [53]. It is an important tool for monitoring the response to treatment and is recommended 3–4 months after initiation of medication or a change in therapy [54].

- Pulmonary arterial systolic pressure (PASP) = RV-RA gradient + right atrium pressure.

$$\text{Mean PAP} = 0.6 * \text{SPAP} + 2\,\text{mmHg} \tag{1}$$

Generally PASP > 38 mmHg detected by echo which suggest mean PAS > 25 mmHg [54].

Right atrial pressure was estimated by the response of the inferior vena cava diameter to inspiration, the value is considered to be about 5–10 mmHg [26].

- Mean pulmonary artery pressure range:

Normal: (12–16) range, mild: (≥25–40) range, moderate: (≥40–<55) range, sever: (≥55) range.

- Systolic pulmonary artery pressure categorized as [55]:

Mild: 40–45 mmHg, moderate: 46–60 mmHg, severe: exceed 60 mmHg.

- The following associated with pulmonary hypertension in echo [56]:

    1. RV, RA hypertrophy or dilatation

    2. Dilated pulmonary artery

9.2 Electrocardiogram (ECG)

The ECG may proffer suggestive or supportive evidence of PAH by finding right ventricular hypertrophy and strain and right atrial dilation. Right ventricular hypertrophy is present in 87% on ECG and in 79% of patients with idiopathic PAH have right axis deviation [57]. The ECG has inadequate sensitivity (55%) and specificity (70%) to be a screening instrument for detecting significant PAH [58].

9.3 Chest radiography (CXR)

Chest radiography is usually performed as the earlier imaging study and may show features of PHT. The classic radiographic findings are evident only late in the disease process [59] including central pulmonary arterial dilatation, variable peripheral lung field that contrasts with pruning of the peripheral blood vessels associated with decreased pulmonary blood flow resulting from increment in pulmonary vascular resistance and the lung became oligemic progressively [45]. In advanced cases, right atrial and ventricular enlargement might appeared and it progressed [57].

The degree of hypertension in any given patient does not correlated with the extent of X-ray abnormalities and normal chest radiograph does not exclude PAH [49].

9.4 Right heart catheterization and vasoreactivity

Right heart catheterization (RHC) remains the gold standard for the diagnosis of pulmonary hypertension, if the patients are prospect for treatment by clinical and transthoracic echocardiographic assessment [26]. He should undergo a confirmatory catheterization of the heart [54]. It is needed to confirm the diagnosis, assessment of the severity of hemodynamic impairment (right atrial pressure, pulmonary artery pressure, pulmonary vascular resistance, to carry out vasoreactivity testing of the pulmonary circulation in selected patients to acute vasodilator testing and to rule out subtle congenital heart disease like pulmonary vein disease [53]. This was needed to performed well expert centers, the techniques have low morbidity (1.1%) and mortality (0.055%) rate [60].

This involves passing a thin flexible tube (catheter) into the right side of the heart which usually passing through vessels of the groin or arm [61].

9.5 High-resolution computed tomography, contrast-enhanced computed tomography

It is a widely available device that can give good information on abnormalities of vascular, cardiac, parenchymal and mediastinal. It can suggest the diagnosis of PAH (PA or RV enlargement), identify the cause of PAH such as lung disease, provide evidences as to the form of PAH and as well provide prognostic information [62].

9.6 Cardiac magnetic resonance imaging

Is an accurate in the assessment of right ventricular size, morphology and function because it is noninvasive so can be used to follow patients regularly, it is an important advantage upon invasive right heart catheterization because measures of RV function have been revealed to be prognostic of long term outcomes in the disease, also provide important information about a patient's disease course and response to treatment by changes in RV function [63].

9.7 Abdominal ultrasound scan

Liver cirrhosis or portal hypertension can be surely excluded through abdominal ultrasound scan and using of contrast agents can improve the diagnosis [57].

10. Treatment

There are no specific treatment guidelines [24].

There are two main parts of management:

1. Predominantly hemoglobinopathy specific treatment and

2. PAH-specific treatment.

It is important to maximize specific treatment of primary hemoglobinopathy, which includes blood transfusion, iron chelation and hydroxyurea, while PAH-specific therapy consist of anticoagulation, diuretics, digoxin, oxygen, and PAH specific vasodilator agents [24].

10.1 Transfusion therapy

Treatment of β-thalassemia depend on an accurate transfusion strategy, particularly in more severe thalassemia major, so to control chronic hemolysis. A chronic transfusion protocol, plus iron chelating therapy to prevent iron overload, is believed to prevent or even ameliorate PAH in these patients [64]. The higher level of blood transfusion may imply more severe disease in patients with PHT, additionally, more chelation therapy used in the PHT group likely reflects more aggressive treatment [33]. So, soon after the diagnosis pulmonary hypertension, regular blood transfusion and appropriate iron chelation therapy can avoid this complication [35].

β-Thalassemia major patients need transfusions throughout life to reach the target Hb level in the range (9–10 g/dl) and to support normal growth, chelation therapy is being used to prevent toxic effects of iron overload, it is generally recommended after twenty units of packed cell transfusion or when serum ferritin exceeds (1000 μg/L), this chelating agents can form complexes with iron and promote its excretion which can clear plasma non transferring bound iron, so remove excess iron from cells and make body iron at safe levels [13].

10.2 Hydroxyurea therapy

Hydroxyurea is a useful treatment, it helps to reduce hemolysis, rise Hb F production (modifies the defective hemoglobin synthesis), so prevents hemolysis and induces nitric oxide in endothelial cells, improves the clinical symptoms, so decreased requirement for blood transfusion, and prevent the acute episodes that exacerbate PAH and potentially reduce overall mortality [24] also has been shown to protect from the development of PA [65]. Additionally, hydroxyurea reduces thrombocytosis, therefore preventing hypercoagulability [34].

Hydroxyurea is an antimetabolite s-phase specific drug that reversibly inhibit ribonucleoside diphosphate reductase enzyme which catalyzes the conversion of ribonucleotides to deoxyribonucleotides which is an important step in biosynthesis of DNA, therefore prevent progression of cell. Also it induces gamma chain globin in human erythroid cell to produce HbF which distributed in RBC [66].

It is used in PAH to reduce the level of circulating immature bone marrow cell and interrupt the abnormal narrowing and occlusion of pulmonary arteries [66].

Dose 10–15 mg/kg/day with gradually increased the dose in step of 2.5–5 mg/kg/day to reach a usual dose 15–30 mg/kg/day (the maximum dose is 35 mg/kg/day) [66].

10.3 L-Carnitine

Mitochondrial dysfunction recently has gained further much attention in the pathophysiology of PAH associated with other diseases, like chronic hemolytic anemia. L-Carnitine is important for the fatty acid oxidation and the normal mitochondrial function [67]. It enhances myocardial function and could increase nitric oxide in plasma which is considered as potent vasodilator that oppose PAH [67] leading to a significant reduction in systolic PAP [22] about 10 mmHg after 3 months of therapy and it is given in a dose of 50 mg/kg/day for 3 months [19].

10.4 Supportive care

All thalassemia patients with PAH, supportive measures should be used immediately. Oxygen therapy should be given to avoid hypoxia which prevent the harmful effects of hypoxia causing vasoconstriction in the pulmonary blood vessels [24]. Physical therapy and lifestyle modifications are often doing to improve symptoms and signs. Immunization is recommended routinely with pneumococcal and influenza vaccines annually [19].

Also, we recommended that we should not use decongestant drugs like pseudoephedrine or other stimulant type medication and avoid use oral contraceptive agent.

Vasoreactivity testing: cardiac catheterization with acute vasodilator drug is essential before select main therapy in children [53].

10.5 PH-specific therapies

These are include

10.5.1 Phospho-diesterase type 5 inhibitors

The cyclic guanosine monophosphate (cGMP) pathway is an essential target of endothelium-derived nitric oxide, an important vasodilator agent. Phospho diestrase type 5 (PDE5) rapidly hydrolyze to cGMP, inactivating its vasodilatory effect. PDE5 inhibitors have been developed that prolong the effect of cGMP through inhibiting its degradation [19].

Sildenafil citrate is a selective, potent inhibitor of the cyclic guanosine monophosphate specific phosphodiesterase-5, which stimulates selective relaxation smooth muscle relaxation in pulmonary vaessels and has been used successfully in the treatment of PAH class II–IV in adult. Prolong therapy with it has been appear to produce a significant and persistent reduction in PAH in patients with thalassemia and sickle thalassemia [68].

10.5.2 Endothelin receptor antagonists

Bosentan has a dual (endothelin) ET receptor antagonist, which reduced pulmonary artery pressure and resistance and improves exercise tolerance in pulmonary hypertension above 12 years [63] class III and IV patients and recently shown beneficial effects in class II patients [53].

Liver function test need follow up regularly as increased in 12% in adult and only 3.5% in children [53].

10.5.3 Soluble guanylate cyclase stimulator

Riociguat, acts directly on a soluble guanylate cyclase stimulating the enzyme which converts guanosine triphosphate to cyclic guanosine monophosphate, increasing sensitivity to low nitric oxide levels leading to vasorelaxation [69]. It has been beneficial in pulmonary arterial hypertension treatment [70] in adult patients with group 4 after surgical treatment or inoperable chronic thromboembolic pulmonary hypertension to optimize exercise capacity as patients with thalassemia syndrome have higher risk of thromboembolic disease leading to pulmonary pressure overload [19].

Prostacyclins: epoprostenol—prostacylin, was considered a key standard for treatment of severe cases.

Epoprostenol given as continuous infusion, it antithrombotic agent that inhibits hypoxic pulmonary vasoconstriction and antiproliferative properties [33].

It improves both the survival and the symptoms of idiopathic form of PHT compared to conventional therapy, it prolongs survival from 35–63% at 3 years [33].

11. Conclusion

Pulmonary hypertension has commonly identified as life threatening for many chronic illness like the hemolytic anemia. The pathophysiology is complex, thus, a comprehensive assessment is imperative when identify appropriate therapy by well control anemia by blood transfusion resulting less risk of pulmonary hypertension. There was no standardization therapy to these patients, therefore we must continue to rely on stories evidence and small cases scenario.

We should make effort for sustained research to improve best practice in vulnerable individuals. Consequently we recommended for annual monitoring by echocardiogram and cardiac catheterization which indicated for persistent elevation of tricuspid regurgitation jet velocity and advise the family to close follow up to maintain Hb above 9 g/dl and ferritin level of less than 1000 ng/ml.

Conflict of interest

There is no conflict of interest to be announced.

Author details

Ahmed Shemran Mutlaq Alwataify[1,2,3*], Sabih Salih Alfatlawy[1] and Yahia Abid Alshahid Altufaily[1,2]

1 Faculty of Medicine, Department of Pediatric, Babylon University, Babylon, Iraq

2 Babylon Maternity and Children Hospital, Babylon, Iraq

3 Babylon Hereditary Blood Disease Center, Babylon, Iraq

*Address all correspondence to: dr.ahmedshemranwataify@gmail.com

References

[1] Rachmilewitz EA, Giardina PJ. How I treat thalassemia. Blood. 2011 Sep 29;**118**(13):3479-3488

[2] Muncie Jr HL, Campbell JS. Alpha and beta thalassemia. American Family Physician. 2009;**80**(4):339-344

[3] Kohne E. Hemoglobinopathies: Clinical manifestations, diagnosis, and treatment. Deutsches Ärzteblatt International. 2011;**108**(31-32):532

[4] Modell B, Darlison M. Global epidemiology of haemoglobin disorders and derived service indicators. Bulletin of the World Health Organization. 2008;**86**:480-487

[5] Farmakis D, Aesso A. Pulmonary hypertension associated with hemoglobinopathies: https;//10.1161/CIRCULATIONAHA.110.988089/Circulation:2011;123;1227-1232.

[6] Taher A, Isma'eel H, Cappellini MD. Thalassemia intermedia: Revisited. Blood Cells, Molecules, and Diseases. 2006;**37**(1):12-20

[7] Musallam KM, Taher AT, Rachmilewitz EA. β-Thalassemia intermedia: a clinical perspective. Cold Spring Harbor Perspectives in Medicine. 2012;**2**(7):a013482

[8] Fathi A, Amani F, Saki M. Prevalence of pul.HT in patients with thalassemia. Pediatric Dimensions. 2016;1(4):95-97. DOI: 10.15761/PD.1000121

[9] Rund D, Rachmilewitz E. β-Thalassemia. New England Journal of Medicine. 2005;**353**(11):1135-1146

[10] Pootrakul P, Sirankapracha P, Hemsorach S, Moungsub W, Kumbunlue R, Piangitjagum A, et al. A correlation of erythrokinetics, ineffective erythropoiesis, and erythroid precursor apoptosis in Thai patients with thalassemia. American Society of Hematology. 2000;**96**(7):2606-2612

[11] Vlahos AP, Koutsouka FP, Papamichael ND, Makis A, Baltogiannis GG, Athanasiou E, et al. Determinants of pulmonary hypertension in patients with beta-thalassemia major and normal ventricular function. Acta Haematologica. 2012;**128**(2):124-129

[12] Galanello R, Origa R. Beta-thalassemia. Orphanet Journal of Rare Diseases. 2010;**5**(1):11

[13] Boddu A, Kumble A, Mahalingam S, Baliga BS, Achappa B. Pulmonary dysfunction in children with beta thalassemia major in relation with iron overload-a cross sectional hospital based study. Asian Journal of Medical Sciences. 2015;**6**(5):47-50

[14] Auger D, Pennell DJ. Cardiac complications in thalassemia major. Annals of the New York Academy of Sciences. 2016;**1368**(1):56-64

[15] Hershko C. Pathogenesis and management of iron toxicity in thalassemia. Annals of the New York Academy of Sciences. 2010;**1202**(1):1-9

[16] Alshemmari ZH, Essam J, Zwaini A, AlJanabi MK. B-thalassemia major in Ramadi. Journal of the Faculty of Medicine. 2005;**47**(2):109-113

[17] Nienhuis AW, Nathan DG. Pathophysiology and clinical manifestations of the β-thalassemias. Cold Spring Harbor Perspectives in Medicine 2012;2(12):a011726.

[18] Kato GJ, Onyekwere OC, Gladwin MT. Pulmonary hypertension in sickle cell disease: relevance to children. Pediatric Hematology and Oncology. 2007;**24**(3):159-170

[19] Fraidenburg DR, Machado RF. Pulmonary hypertension associated with thalassemia syndromes. Annals of the New York Academy of Sciences. 2016;**1368**(1):127-139

[20] Kontoghiorghes GJ, Kolnagou A. Deferiprone versus desferrioxamine in thalassaemia, and T2* validation and utility. The Lancet. 2003;**361**(9352):184

[21] Galie N, Torbicki A, Barst R, Dartevelle P, Haworth S, Higenbottam T, et al. Guidelines on diagnosis and treatment of pulmonary arterial hypertension. European Heart Journal. 2004 Dec 1;**25**(24):2243-2278

[22] Frank BS, Ivy DD. Diagnosis, evaluation and treatment of pulmonary arterial hypertension in children. Children. 2018;**5**(4):44

[23] Azami M, Sufi Nia A, YektaKooshali MH, Nikpay S, Madmoli Y, Malekshahi M, et al. Prevalence and risk factors of pulmonary arterial hypertension in thalassemia major patients of Ilam, 2014. Evidence Based Care. 2017;**6**(4): 74-78

[24] Saleemi S. Saudi guidelines on the diagnosis and treatment of pulmonary hypertension: Pulmonary hypertension associated with hemolytic anemia. Annals of Thoracic Medicine. 2014; **9**(Suppl 1):S67

[25] Chueamuangphan N, Chuncharunee S, Atichartakarn V, Likittanasombat K, Sriwattanakomen O. Pulmonary arterial hypertension in B-thalassemia. Journal of Hematology and Transfusion Medicine. 2009;**19**(2): 101-108

[26] Schannwell CM, Steiner S, Strauer B. Diagnostics in pulmonary hypertension. Journal of Physiology and Pharmacology. 2007;**58**(5):591-602

[27] Hoeper MM, Bogaard HJ, Condliffe R, Frantz R, Khanna D, Kurzyna M, et al. Definitions and diagnosis of pulmonary hypertension. Journal of the American College of Cardiology. 2013;**62**(25 Supplement): D42-D50

[28] Simonneau G, Gatzoulis MA, Adatia I, Celermajer D, Denton C, Ghofrani A, et al. Updated clinical classification of pulmonary hypertension. Journal of the American College of Cardiology. 2013; **62**(25 Supplement):D34-D41

[29] Chueamuangphan N, Wongtheptien W, Nawarawong W, Sukornthasarn A, Chuncharunee S, Tawichasri C, et al. Clinical indicators for pulmonary arterial hypertension in thalassemia. Journal of the Medical Association of Thailand. 2012;**95**(1):16

[30] Anthi A, Orfanos SE, Armaganidis A. Pulmonary hypertension in β thalassaemia. The Lancet Respiratory Medicine. 2013;**1**(6):488-496

[31] Walford G, Loscalzo J. Nitric oxide in vascular biology. Journal of Thrombosis and Haemostasis. 2003; **1**(10):2112-2118

[32] Adedeji MO, Cespedes J, Allen K, Subramony C, Hughson MD. Pulmonary thrombotic arteriopathy in patients with sickle cell disease. Archives of Pathology & Laboratory Medicine. 2001;**125**(11): 1436-1441

[33] Nonlawan C, Suporn C, Vichai A, Khanchit L, Orapan S. Pulmonary hypertension in B thalassemia. Journal of Hematology & Transfusion. 2008;**19**: 101-108

[34] Farmakis D, Aessopos A. Pulmonary hypertension associated with hemoglobinopathies: Prevalent but overlooked. Circulation. 2011;**123**(11): 1227-1232

[35] Moghaddam HM, Badiei Z, Eftekhari K, Shakeri R, Farhangi H.

Prevalence of pulmonary hypertension in patients with thalassemia intermedia in 2009: A single center's experience. Electronic Physician. 2015;7(3):1102

[36] Farmaki K, Tzoumari I, Pappa C, Chouliaras G, Berdoukas V. Normalisation of total body iron load with very intensive combined chelation reverses cardiac and endocrine complications of thalassaemia major. British Journal of Haematology. 2010; **148**(3):466-475

[37] Wood JC. Cardiac complications in thalassemia major. Hemoglobin. 2009;**33**(Suppl 1):S81-S86

[38] Humbert M, Sitbon O, Chaouat A, Bertocchi M, Habib G, Gressin V, et al. Pulmonary arterial hypertension in France. American Journal of Respiratory and Critical Care Medicine. 2006; **173**(9):1023-1030

[39] Nonlawan C., Wattana W, Apichard S, Suporn C., Chamaiporn T and Jayanton P.:Chueamuangphan N, Wongtheptien W, Nawarawong W, Sukornthasarn A, Chuncharunee S, Tawichasri C, Patumanond J Clinical indicators for pulmonary arterial hypertension in thalassemia: Journal of the Medical Association of Thailand. Thai 2012;95(1): 16-21.

[40] Mardan RH, Radi HY, Alwataify AS, Altufaily YA. The incidence of pulmonary HT among thalassemia patients in Babylon hereditary blood disease center in Babylon governorate/ Iraq. Biochemical and Cellular Archives: the paper was accepted to be appeared in vol. 21.No. p. 2531-2538. (Issue 2 october 2021).

[41] Hoeper MM, Jost N, Frank H, Peer F, Fabel H. Pulmonary hypertension after splenectomy. Annals of Internal Medicine. 1999;**130**(6):506-509

[42] Gladwin MT, Vichinsky E. Pulmonary complications of sickle cell disease. New England Journal of Medicine. 2008;**359**(21):2254-2265

[43] Frost A, Badesch D, Gibbs JS, Gopalan D, Khanna D, Manes A, et al. Diagnosis of pulmonary hypertension. European Respiratory Journal. 2019; **53**(1):1-12

[44] Ambrusko SJ. Pulmonary hypertension in children with hemolytic disorders. Progress in Pediatric Cardiology. 2020;**56**:101194

[45] Rosenzeig EB, Fein JA, Humpl T, Ivy DD. Pulmonary hypertension in children: Diagnostic work up and challenges. Progress in Pediatric Cardiology. 2009;**27**(1):4-11. DOI: 10.1016/j.ppedcard.2009.09.003

[46] Bogaard HJ, Abe K, Noordegraaf AV, Voelkel NF. The right ventricle under pressure: Cellular and molecular mechanisms of right-heart failure in pulmonary hypertension. Chest. 2009;**135**(3):794-804

[47] Bazan IS, Fares WH. Hypercoagulability in pulmonary hypertension. Clinics in Chest Medicine. 2018;**39**(3):595-603

[48] Olsson KM, Nickel NP, Tongers J, Hoeper MM. Atrial flutter and fibrillation in patients with pulmonary hypertension. International Journal of Cardiology. 2013;**167**(5):2300-2305

[49] Galiè N, Humbert M, Vachiery JL, Gibbs S, Lang I, Torbicki A, et al. ESC/ ERS: Guidelines for the diagnosis and treatment of pulmonary hypertension. European Respiratory Journal. 2015, 2015;**46**(4):903-975

[50] Meloni A, Detterich J, Pepe A, Harmatz P, Coates TD, Wood JC. Pulmonary hypertension in well-transfused thalassemia major patients.

Blood Cells, Molecules, and Diseases. 2015;**54**(2):189-194

[51] Chemla D, Castelain V, Humbert M, Hébert JL, Simonneau G, Lecarpentier Y, et al. New formula for predicting mean pulmonary artery pressure using systolic pulmonary artery pressure. Chest. 2004;**126**(4): 1313-1317

[52] Rosenzweig EB, Abman SH, Adatia I, Beghetti M, Bonnet D, Haworth S, et al. Paediatric pulmonary arterial hypertension: Updates on definition, classification, diagnostics and management. European Respiratory Journal. 2019;**53**(1):1801916

[53] Ivy D. Pulmonary hypertension in children. Cardiology Clinics. 2016; **34**(3):451-471. DOI: 10. 1016/j.ccl2016. 04.005

[54] Habib G, Torbicki A. The role of echocardiography in the diagnosis and management of patients with pulmonary hypertension. European Respiratory Review. 2010;**19**(118): 288-299

[55] Scott DS, Bernard B. Essential Echocardiography: A Practical Guide with DVD. Springer Science & Business Media; 2007

[56] Augustine DX, Coates-Bradshaw LD, Willis J, Harkness A, Ring L, Grapsa J, et al. Echocardiographic assessment of pulmonary hypertension: a guideline protocol from the British Society of Echocardiography. Echo Research and Practice. 2018;**5**(3):G11-G24

[57] Galiè N, Manes A, Branzi A. Evaluation of pulmonary arterial hypertension. Current Opinion in Cardiology. 2004;**19**(6):575-581

[58] Ahearn GS, Tapson VF, Rebeiz A, Greenfield JC Jr. Electrocardiography to define clinical status in primary pulmonary hypertension and pulmonary arterial hypertension secondary to collagen vascular disease. Chest. 2002;**122**(2):524-527

[59] Tunariu N, Gibbs SJ, Win Z, Gin-Sing W, Graham A, Gishen P, et al. Ventilation–perfusion scintigraphy is more sensitive than multidetector CTPA in detecting chronic thromboembolic pulmonary disease as a treatable cause of pulmonary hypertension. Journal of Nuclear Medicine. 2007;**48**(5):680-684

[60] Hoeper MM, Lee SH, Voswinckel R, Palazzini M, Jais X, Marinelli A, et al. Complications of right heart catheterization procedures in patients with pulmonary hypertension in experienced centers. Journal of the American College of Cardiology. 2006;**48**(12):2546-2552

[61] Morris C. Pulmonary hypertension in thalassemia. 2010

[62] Rajaram S, Swift AJ, Condliffe R, Johns C, Elliot CA, Hill C, et al. CT features of pulmonary arterial hypertension and its major subtypes: a systematic CT evaluation of 292 patients from the ASPIRE registry. Thorax. 2015;**70**(4):382-387

[63] Peacock AJ, Noordegraaf AV. Cardiac magnetic resonance imaging in pulmonary arterial hypertension. European Respiratory Review. 2013; **22**(130):526-534

[64] Aessopos A, Farmakis D, Hatziliami A, Fragodimitri C, Karabatsos F, Joussef J, et al. Cardiac status in well treated patients with thalassemia major. European Journal of Haematology. 2004;**73**(5):359-366

[65] Karimi M, Borzouee M, Mehrabani A, Cohan N. Echocardiographic finding in beta thalassemia intermedia and major:

Absence of pulmonary hypertension following hydroxyurea treatment in beta-thalassemia intermedia. European Journal of Haematology. 2009;**82**(3): 213-218

[66] Nirman Y, Anuga P. And Sachith M Comprehensive review of hydroxyurea for B-Hbpathies: The role revisited during COVID-19 pandemic.Orphanet Journal of Rare Diseases 2021;16;114

[67] El-Beshlawy A, Youssry I, El-Saidi S, El Accaoui R, Mansi Y, Makhlouf A, et al. Pulmonary hypertension in β-thalassemia major and the role of L-carnitine therapy. Pediatric Hematology and Oncology. 2008;**25**(8): 734-743

[68] Derchi G, Forni GL, Formisano F, Cappellini MD, Galanello R, D'Ascola G, et al. Efficacy and safety of sildenafil in the treatment of severe pulmonary hypertension in patients with hemoglobinopathies. Haematologica. 2005;**90**(4):452-458

[69] Grimminger F, Weimann G, Frey R, Voswinckel R, Thamm M, Bölkow D, et al. First acute haemodynamic study of soluble guanylate cyclase stimulator riociguat in pulmonary hypertension. European Respiratory Journal. 2009;**33**(4):785-792

[70] Ghofrani HA, Galiè N, Grimminger F, Grünig E, Humbert M, Jing ZC, et al. Riociguat for the treatment of pulmonary arterial hypertension. The New England Journal of Medicine. 2013;**369**:330-340

Chapter 9

Challenges of Hepatitis C Virus Treatment in Thalassemia

Iman El-Baraky

Abstract

Thalassemic patients, especially in limited resources settings, are prone to multi-transfusion acquired Hepatitis C virus (HCV). After the discovery of direct acting antivirals (DAAs), many programs were designed to achieve HCV eradication both on the macro-elimination and micro-elimination axes. Thalassemic patients are good candidates to be addressed by a unique HCV micro-elimination model since they face some challenges during their treatment journey. Some of these challenges are the young age at infection, frequent blood transfusion, polypharmacy, drug–drug interactions, pharmacokinetic considerations and the risk of reinfection. The available data of success rates of HCV cure in thalassemic patients alert that the success rate in thalassemic patients might be lower than that reported in general population. These factors make HCV micro-elimination model, a hurdle towards the 2030 world health organisation (WHO) HCV eradication plan.

Keywords: Thalassemia, Hepatitis C, Direct acting antivirals, challenges, micro-elimination

1. Introduction

Haematologic diseases include any disease affecting the blood or blood forming organs such as bone marrow. Haematologic diseases include haemoglobinopathies, malignancies, different types of anaemia and other less common diseases. Thalassemia is one of the most common haemoglobinopathies [1].

Hepatitis C virus (HCV) is a liver disease that can cause both acute and chronic infections. Chronic HCV infection can lead to liver cirrhosis or hepatocellular carcinoma (HCC). If liver cirrhosis became decompensated, no cure is available except liver transplantation [2].

Thalassemia patients have high prevalence of HCV and are more susceptible to its complications. Some challenges face the health care givers while developing their therapeutic plans for thalassemia patients [3].

2. Thalassemia

Thalassemia is a congenital disease where haemoglobin is malformed resulting in premature haemolysis. Thalassemia is classified into alfa thalassemia and beta thalassemia according to the affected globin chain. Thalassemia is also denoted a grade according to its severity into trait, carrier, intermedia, or major where major is the most severe case [1].

Thalassemia patients suffer from premature haemolysis, leading to increased Red blood cells (RBCs) turnover, ineffective erythropoiesis, and varying degrees of anaemia [4]. Thalassemia can be cured by bone marrow transplantation. If bone marrow transplantation is not feasible, thalassemia can be managed by regular blood transfusion along with other supportive therapies such as iron chelation [1].

2.1 Thalassemia and HCV

Thalassemia patients are at risk of multi-transfusion HCV infection specially in limited resources settings. Due to the relatively young age at infection, some thalassemia patients miss early diagnosis and treatment.

2.1.1 HCV discovery and genotypes

HCV is positive-stranded ribonucleic acid (RNA) enveloped virus which was first discovered in 1989. HCV genome is highly variable. Therefore, HCV has been classified into six genotypes (GTs) numbered from one to six. Different GTs show different response to treatment. Hence, treatment is genotype (GT) based. HCV GTs has numerous sub-genotypes which are also different with regard to response to treatment [2, 5].

The geographic distribution of HCV GTs show that GT-1 is the most prevalent GT worldwide followed by genotype 3. They account for about 46.2% and 30.1% of all HCV infections, respectively. On one hand, some countries have diverse HCV genotypic distribution such as China. On the other hand, in some countries, such as Egypt, more than 90% of HCV infections are caused by single genotype which is GT-4 [2, 6, 7].

2.1.2 HCV prevalence

Globally, between 130 and 150 million people had chronic HCV infection and 700,000 people died due to HCV related complications in 2013 [2, 8]. HCV global paediatric burden was estimated to be 3.5 million children or adolescents [9, 10]. Some countries such as Egypt suffers from an endemic HCV prevalence. HCV prevalence in Egypt was estimated to be the highest worldwide in 2015 where viraemic prevalence was estimated to be 6.3% in adults 1.1% in children aged 10–19 years [7, 11].

2.1.3 Mode of transmission of HCV among thalassemia patients

HCV infection is blood borne. The main source of HCV infection in thalassemia patients is unscreened blood transfusions. This does not preclude that unsafe injection practices, health-care-associated transmission and renal dialysis are among the other causes of HCV transmission in thalassemia patients. According to a previous World Health Organisation (WHO) report on blood safety, some countries do not properly screen blood transfusions for blood borne viruses on routine basis. Hence, the WHO developed its guidance on phlebotomy [2, 12, 13]. Prior to the implementation of the WHO guidance, multi-transfusion associated HCV acquisition was a common mode of transmission among multi-transfusers such as thalassemia patients specially of major type. HCV prevalence by HCV antibody test widely vary from 4–85% according to different reports [14].

Less common routes of HCV transmission among thalassemia patients are mother-to-child, also known as vertical transmission, sexual transmission, and intranasal drug use [2].

2.1.4 Thalassemia consequences that interfere with HCV treatment

Because of premature haemolysis and ineffective erythropoiesis, that thalassemia patients suffer, the core of their supportive care is regular blood transfusion. The major consequences of multi-transfusion are iron overload, oxidative stress, splenomegaly, liver fibrosis and blood borne infections.

Thalassemia induced haemolysis results in chronic anaemia. Severe anaemia is compensated by hyperdynamics. Hyperdynamic results in an initial increase in the cardiac index and blood flow to vital organs such as liver and kidney. However, over long term, thalassemia patients develop heart failure and renal impairment [15–19].

Iron overload can negatively impact the patients' cardiovascular and endocrine systems leading to heart failure, osteoporosis and other diseases [20]. Excessive haemolysis leads to splenomegaly which in turn leads to increased transfusion requirements and difficult control of iron overload. Therefore, many beta-thalassemia patients who suffer severe anaemia are splenectomised [21].

Among the complications of thalassemia are blood borne infections especially in low to intermediate income countries that do not implement rigorous blood products screening systems [2]. Multi-transfusion related HCV is one of the blood borne infections thalassemia patients suffer. According to the WHO Regional Office for the Eastern Mediterranean (WHO-EMRO), 11–69% of Beta-thalassemia major patients suffer from chronic HCV in the aforementioned region [22].

2.1.5 HCV prognosis in thalassemia

The early phase of HCV prognosis in thalassemia patients does not differ secondary to their disease nature. HCV infection manifest in thalassemia patients as acute or chronic hepatitis. 15–45% of cases show spontaneous clearance of untreated acute HCV infection within six months of infection similar to patients without thalassemia. The remaining 55–85% develop chronic HCV which is usually asymptomatic. If left untreated, chronic HCV infection causes liver cirrhosis and consequently liver failure. Liver cirrhosis is associated with high risk of hepatocellular carcinoma (HCC). This normal course of chronic HCV prognosis is aggravated by thalassemia since the pathophysiology of thalassemia include iron overload and consequently, increased oxidative stress that result in necro-inflammatory responses aggravating the course of HCV induced liver fibrosis and eventually cirrhosis. Thalassemia could induce liver fibrosis even on absence of HCV infection due to the iron deposits and iron overload induced oxidative stress. Liver fibrosis progress in thalassemia patients even in absence of HCV infection [2, 23].

2.1.6 HCV treatment

2.1.6.1 HVC treatment development

For decades, there was no available treatment for HCV until the discovery of the role interferon-alpha in HCV immune response. Interferon-alpha-2b/ribavirin for 48 weeks were the standard treatment for more other decades where Interferon was injected three times a week and ribavirin was orally administered daily.

Pegylated interferon-alfa-2b/ribavirin regimen was the next step in the development of interferon-based regimens for the treatment of HCV, where pegylation allowed once weekly injection of interferon. Pegylated interferon/ribavirin regimen was poorly tolerated, and its cure rates were between 40% and 65% depending mainly on the HCV GT and cirrhosis status. HCV cure is measured by sustained virologic response twelve weeks post treatment (SVR12). Pegylated interferon/ribavirin regimen was sometimes associated with life-threatening adverse reactions [2, 8].

The introduction of direct acting antivirals (DAAs) comprised a revolutionary step in HCV treatment. DAAs directly inhibited the replication cycle of HCV through targeting three important regions within the HCV genome. These three regions are the non-structural (NS) NS3/4A protease, NS5A and NS5B RNA-dependent polymerase. DAAs achieved higher success rate than interferon-based regimens. DAAs based therapy has a success rate of about 90% after 12 weeks therapy. Other advantages of DAAs based regimen are the shorter treatment duration which range from 8 to 24 weeks, oral administration leading to better compliance and more favourable safety profile. DAAs based regimens are combination therapies. Individual DAAs based regimens vary in their therapeutic efficacy, genotypic efficacy, and safety profile [2, 8].

The first-generation DAAs were protease inhibitors. They were co-administered with interferon and ribavirin according to the WHO guidelines in 2014. However, they were only effective against GT-1 infection. In addition, they had frequent and sometime severe side-effects. Therefore, the WHO no longer recommends first generation DAAs.

Second-generation DAAs have higher SVR12 rates and better safety than the first generation. The strongest advantage of the second generation DAAs is that they can be used in combinations obviating the need for interferon and ribavirin. Interferon-free combinations of two or three second generation DAAs have demonstrated excellent efficacy in general, although cure rates among certain patient subgroups were lower [2, 8].

The WHO currently recommends that, all HCV infected patients to be treated with DAAs based therapy, except those having GT-5 or 6 infection and cirrhotic GT-3 patients in whom interferon-based therapy can still be used as an alternative regimen. DAAs are tremendously developing with the aim of improved efficacy and safety profiles [2].

Some DAAs combination showed approximately 100% cure rate in adults such as sofosbuvir (SOF)/simeprevir (SMV) combination with or without ribavirin for 24 weeks, SOF/daclatasvir (DAC) combination with or without ribavirin for 24 weeks and ombitasvir/paritaprevir/ritonavir combination with either dasabuvir or ribavirin for 12 weeks [2, 8].

In children, until recently, the only food and drug administration (FDA) approved regimens were interferon based [24]. The treatment success rates and adverse effects of interferon-based therapy among children were similar to those of adults [25].

Till January 2017, six DAAs combined into four regimens with ribavirin were in clinical trials for paediatric patients. These regimens are SOF/ribavirin, SOF/ledipasvir (LDV) and paritaprevir/ombitasvir/ritonavir ± ribavirin/dasabuvir. The duration of therapy ranged from 12 to 24 weeks [24]. SOF/LDV for 12 weeks proved to be 98% effective in adolescents with genotype-1 infection with favourable safety profile [26]. Consequently, in April 2017, the FDA approved SOF/LDV and SOF/ribavirin for HCV treatment in paediatric patients aged 12–17 years [27]. SOF/LDV had 98–100% SVR12 rate in adolescents [28–33]. SOF/LDV SVR12 rate sometimes differs in special sub-populations where its SVR12 rate in adolescent beta-thalassemia major patients was previously reported to be 89% (95% confidence interval (CI) 74–100%) [34]. Later in 2017, a clinical trial of glecaprevir (GLC)/pibrentasvir (PBR) for 8–16 weeks in adolescents were launched. Glecaprevir/pibrentasvir was approved for use in adolescents in April 2019 [35]. After about one year, SOF/velpatasvir (VLP) was approved in March 2020 [36].

Until the current date, only four DAAS based regimens were approved in paediatric patients namely, SOF/ribavirin, SOF/LDV, GLC/PBR and SOF/VLP [27, 35, 36]. In April 2017, SOF/ribavirin and SOF/LDV were approved in adolescents weighing at least 35 Kg. Afterwards, the FDA approved the pan-genotypic GLC/PBR in adolescents weighing at least 45 Kg and approved SOF/LDV use in children starting from the age of 3 years. The last milestone in the DAAs development process in paediatric patients was the approval of the pan genotypic SOF/VLP in children starting from the age of six years [27, 35–37].

2.1.6.2 HCV treatment in thalassemia

Before April 2017, the only available HCV treatment for paediatric patients was interferon-alfa 2b/ ribavirin-based regimen. Ribavirin based regimens were of limited applicability in thalassemia patients due to ribavirin haematologic side effects which include haemolysis. Interferon monotherapy had about 30% success rate in terms of sustained virologic response 12 weeks post-treatment (SVR12). Ribavirin use in thalassemia patients lead to higher SVR12 at the expense of increased blood requirements during the treatment period [22].

Before the approval of direct acting antivirals (DAAs) based regimens in April 2017, thalassemia paediatric patients were supported by different hepatoprotective agents to counteract HCV induced liver fibrosis and cirrhosis until the initiation of therapy. Hepatoprotectives were also recommended as an adjunct to DAAs based therapy to stop the progression of liver fibrosis after SVR12 achievement. Hepatoprotective agents include some plant derived compounds such as silymarin, silibinin, and Curcumin, antioxidants, such as N-acetyl cysteine, L-carnitine, and vitamin E and some drugs such as metformin hydrochloride [23, 38, 39].

After adulthood, many DAAs based regimens are available nowadays. HCV treatment in selected based on evidence based medicine [2]. Selection of DAAs based regimen is mainly dependent on the HCG genotype, patient's cirrhosis previous treatment trials. Many regimens achieved SVR12 rated exceeding 90% with acceptable safety level. SVR12 in thalassemia patients after DAAs based regimens were reported to be 80–100%. These regimens include SOF/LDV, SOF/DAC, SOF, SOF/simeprevir (SIM), elbasvir/grazoprevir and ombitasvir/pibrentasvir/glecaprevir [34, 40–46]. The interest in investigating the reasons behind treatment failure among thalassemia patients is growing.

The WHO set a goal and targeted to eliminate HCV by 2030 [10]. To eliminate HCV infection, treatment should be offered and optimised not only in the general population, but also in special sub-populations. Thalassemia patients have an increased risk of multi-transfusion acquired HCV, particularly in resource limited settings, and an increased risk of its hepatic complications. In addition, their congenital disease nature may influence the pharmacokinetics of some drugs [3, 20, 22, 47].

2.1.6.3 Pharmacokinetic considerations in thalassemia

Thalassemia's effect on the pharmacokinetics of drugs might be attributed to haemolysis, hyperdynamics, altered plasma protein binding, splenectomy, blood transfusion or drug–drug interactions [1, 20].

Some anti-HCV prodrugs are activated in the RBCs such as SOF [48]. Consequently, SOF activation might be altered in thalassemia patients.

Thalassemia induced hyperdynamics was found to increase the cardiac index by 60% in beta-thalassemia major patients. Since the hepatic blood flow comprises about 25% of the cardiac output, it was expected to rise consequently [18, 19, 49].

Since the hepatic metabolism of high extraction ratio drugs is flow dependent [50]. Therefore, hyperdynamics-enhanced hepatic blood flow might affect the clearance of high hepatic extraction ratio anti-HCV drugs such as SOF and LDV.

During the hyperdynamic phase, thalassemia patients were also found to go through initial renal hyperfiltration. Renal hyperfiltration might enhance the clearance of renally eliminated drugs such as SOF metabolite GS-331007 [51, 52]. Eventually, thalassemia patients suffer from renal impairment secondary to chronic anaemia, iron overload and iron chelators induced nephrotoxicity [17].

Thalassemia induced iron overload might alter the plasma protein binding of highly plasma protein bound anti-HCV drugs such as LDV, telaprevir, asunaprevir, DAC, ombitasvir, elbasvir, VLP, and dasabuvir that are at least 99% plasma protein bound [53, 54]. Blood transfusion might also alter the steady state of regularly administered drugs.

Moreover, splenectomy was proved to affect the pharmacokinetics of some iron chelating drugs [55]. The effect of splenectomy on anti-HCV drugs needs further research. It has been found that splenectomy does not affect the pharmacokinetics of SOF/LDV [56] but its effect on other anti-HCV drugs needs further investigation.

In conclusion, thalassemia may affect the pharmacokinetics of some anti-HCV drugs such as SOF, SOF metabolite GS-331007 and LDV [56] and some aspects still need further studies such as plasma protein binding alteration, blood transfusion effect on the drugs levels, splenectomy effect and drug–drug interactions.

3. Conclusions

Many challenges still awaiting thalassemia HCV micro-elimination model. More research is still required to rationalise treatment failure.

Conflict of interest

The authors declare no conflict of interest.

Author details

Iman El-Baraky
Faculty of Pharmacy, Cairo University, Cairo, Egypt

*Address all correspondence to: eman.ali@pharma.cu.edu.eg

References

[1] Centres for Disease Control and prevention (CDC), *Thalassemia*. 2017. Available from: https://www.cdc.gov/ncbddd/thalassemia/facts.html. Accessed: 15th of Octocer 2017.

[2] Guidelines for the screening, care and treatment of persons with chronic hepatitis C infection. World Health Organisation (WHO). 2016. Availablle from: http://apps.who.int/iris/bitstream/10665/205035/1/978924 1549615_eng.pdf?ua=1. Accessed: 15th of March 2017.

[3] Abdella, Y., Riedner, G., Hajjeh, R., *et al. Blood transfusion and hepatitis: what does it take to prevent new infections?* Eastern Mediterranean Health Journal. 2018; 24(6):595-597.

[4] Rachmilewitz, E.A., Weizer-Stern, O., Adamsky, K., *et al. Role of iron in inducing oxidative stress in thalassemia: can it be prevented by inhibition of absorption and by antioxidants?* Ann. N. Y. Acad. Sci. 2005; 1054(1):118-123.

[5] Jafri, W., Siddiqui, B. and Awan, S. *HCV-discovery to elimination,"myth or reality"*. J. Hepatoma Res. 2018; 4(54):1-8.

[6] Messina, J.P., Humphreys, I., Flaxman, A., *et al. Global distribution and prevalence of hepatitis C virus genotypes*. Hepatology. 2015; 61(1):77-87.

[7] *Global prevalence and genotype distribution of hepatitis C virus infection in 2015: a modelling study*. Lancet Gastroenterol. Hepatol. 2017; 2(3):161-76.

[8] *Hepatitis C fact sheet*. World health organisation (WHO). 2020. Available from: http://www.who.int/mediacentre/factsheets/fs164/en/. Accessed: 19th of june 2021.

[9] Indolfi, G., Easterbrook, P., Dusheiko, G., *et al. Hepatitis C virus infection in children and adolescents*. Lancet Gastroenterol. Hepatol. 2019; 4:477-487.

[10] *Hepatitis C*. World Health Organisation (WHO). 2020. Available from: https://www.who.int/news-room/fact-sheets/detail/hepatitis-c. Accessed: 19th of June 2021.

[11] Gomaa, A., Allam, N., Elsharkway, A., *et al. Hepatitis C infection in Egypt: prevalence, impact and management strategies*. Hepat. Med. 2017; 9:17-25.

[12] Frank, C., Mohamed, M.K., Strickland, G.T., *et al. The role of parenteral antischistosomal therapy in the spread of hepatitis C virus in Egypt*. The Lancet. 2000; 355(9207):887-891.

[13] WHO guidelines on drawing blood: best practices in phlebotomy. World Health Organisation (WHO). 2010. Availablle from: https://www.euro.who.int/__data/assets/pdf_file/0005/268790/WHO-guidelines-on-drawing-blood-best-practices-in-phlebotomy-Eng.pdf. Accessed: 21st of March 2017.

[14] Mansour, A.K., Aly, R.M., Abdelrazek, S.Y., *et al. Prevalence of HBV and HCV infection among multi-transfused Egyptian thalassemic patients*. Journal of Hematology/oncology stem cell therapy. 2012; 5(1):54-59.

[15] Aessopos, A., Farmakis, D., Tsironi, M., *et al. Hemodynamic assessment of splenomegaly in β-thalassemia patients undergoing splenectomy*. Ann. Hematol. 2004; 83(12):775-778.

[16] Bekhit, O.E., El Dash, H.H. and Ahmed, M.S. *Early detection of kidney dysfunction in Egyptian patients with beta-thalassemia major*. Egyptian Pediatric Association Gazette. 2017; 65(3):85-89.

[17] Demosthenous, C., Eleftheriou, P., Apostolou, C., *et al. β-Thalassemia and*

renal complications. A narrative review of pathophysiologic mechanisms. Integrative Molecular Medicine. 2018; 5(4):1-10.

[18] Kremastinos, D.T., Farmakis, D., Aessopos, A., *et al. β-thalassemia cardiomyopathy: history, present considerations, and future perspectives.* Circ. Heart Fail. 2010; 3(3):451-458.

[19] Pennell, D.J., Udelson, J.E., Arai, A.E., *et al. Cardiovascular function and treatment in β-thalassemia major: a consensus statement from the American Heart Association.* Circulation. 2013; 128(3):281-308.

[20] Malik, S., Syed, S. and Ahmed, N. *Complications in transfusion–dependent patients of ß-thalassemia major: A review.* Pakistan Journal of Medical Sciences. 2009; 25(4):678-682.

[21] Rund, D. and Rachmilewitz, E. *β-Thalassemia.* N. Engl. J. Med. 2005; 353(11):1135-1146.

[22] Daher, H.B. and Sharara, A.I. *Treatment of chronic HCV infection in patients with thalassemia.* Clinical Liver Disease. 2019; 14(6):199-202.

[23] Poustchi, H., Eslami, M., Ostovaneh, M.R., *et al. Transient elastography in hepatitis C virus-infected patients with beta-thalassemia for assessment of fibrosis.* Hepatol. Res. 2013; 43(12):1276-1283.

[24] Schwarz, K.B. and Karnsakul, W. *Treatment of Hepatitis C in Children.* Current Hepatology Reports. 2017; 16(1):18-25.

[25] Hu, J., Doucette, K., Hartling, L., *et al. Treatment of hepatitis C in children: a systematic review.* PLoS One. 2010; 5(7):e11542.

[26] Balistreri, W.F., Murray, K.F., Rosenthal, P., *et al. The safety and effectiveness of ledipasvir- sofosbuvir in adolescents 12 to 17 years old with hepatitis C virus genotype 1 infection.* Hepatology. 2016.

[27] *FDA approves two hepatitis C drugs for pediatric patients.* FDA News Release. Food and Drug Administration (FDA). 2017. Available from: https://www.fda.gov/newsevents/newsroom/pressannouncements/ucm551407.htm. Accessed: 25th of August 2018.

[28] Balistreri, W.F., Murray, K.F., Rosenthal, P., *et al. The safety and effectiveness of ledipasvir- sofosbuvir in adolescents 12-17 years old with hepatitis C virus genotype 1 infection.* Hepatology. 2017; 66(2):371-378.

[29] El-Khayat, H., Kamal, E., El-Sayed, M., *et al. The effectiveness and safety of ledipasvir plus sofosbuvir in adolescents with chronic hepatitis C virus genotype 4 infection: a real-world experience.* Aliment. Pharmacol. Ther. 2018; 47(6):838-844.

[30] El-Araby, H.A., Behairy, B.E., El-Guindi, M.A., *et al. Generic sofosbuvir/ledipasvir for the treatment of genotype 4 chronic hepatitis C in Egyptian children (9-12 years) and adolescents.* Hepatol. Int. 2019; 13(6): 706-714.

[31] Harvoni (sofosbuvir/ledipasvir) tablets, for oral use. Gilead Sciences Inc. 2019. Available from: https://www.accessdata.fda.gov/drugsatfda_docs/label/2019/212477s000lbl.pdf. Accessed: 8th of April 2020.

[32] Smolders, E.J., Jansen, A.M., ter Horst, P.G., *et al. Viral hepatitis C therapy: pharmacokinetic and pharmacodynamic considerations: a 2019 update.* Clin. Pharmacokinet. 2019; 58:1237-1263.

[33] El-Karaksy, H., Mogahed, E.A., Abdullatif, H., *et al. Sustained viral response in Genotype 4 chronic Hepatitis C virus–infected children and adolescents treated with sofosbuvir/ledipasvir.* J.

Pediatr. Gastroenterol. Nutr. 2018; 67(5):626-630.

[34] Nagral, A., Jhaveri, A., Sawant, S., *et al. Treatment of chronic hepatitis C infection with direct acting antivirals in adolescents with thalassemia major*. The Indian Journal of Pediatrics. 2019; 86(2):148-153.

[35] *FDA approves first treatment for all genotypes of hepatitis C in pediatric patients*. FDA news release. Food and Drug Administration (FDA). 2019. Available from: https://www.fda.gov/news-events/press-announcements/fda-approves-first-treatment-all-genotypes-hepatitis-c-pediatric-patients. Accessed: 1st of October 2020.

[36] FDA approves new treatment for pediatric patients with any strain of Hepatitis C. Food and Drug Administration (FDA). 2020. Availablle from: https://www.fda.gov/news-events/press-announcements/fda-approves-new-treatment-pediatric-patients-any-strain-hepatitis-c. Accessed: 8th of June 2020.

[37] *FDA approves label changes for Sovaldi, Harvoni in children with HCV*. FDA news release. Food and Drug Administration (FDA). 2019. Available from: https://www.healio.com/news/hepatology/20190905/fda-approves-label-changes-for-sovaldi-harvoni-in-children-with-hcv. Accessed: 1 October, 2020.

[38] Abdel Monem, M.S., Farid, S.F., Abbassi, M.M., *et al. The potential hepatoprotective effect of metformin in hepatitis C virus-infected adolescent patients with beta thalassemia major: Randomised clinical trial*. J International Journal of Clinical Practice. 2021; 75(6):e14104.

[39] Li, H., Huang, M.-H., Jiang, J.-D., *et al. Hepatitis C: From inflammatory pathogenesis to anti-inflammatory/hepatoprotective therapy*. J World journal of gastroenterology. 2018; 24(47):5297.

[40] Mangia, A., Sarli, R., Gamberini, R., *et al. Randomised clinical trial: sofosbuvir and ledipasvir in patients with transfusion-dependent thalassaemia and HCV genotype 1 or 4 infection*. Aliment. Pharmacol. Ther. 2017; 46(4):424-431.

[41] Sinakos, E., Kountouras, D., Koskinas, J., *et al. Treatment of chronic hepatitis C with direct-acting antivirals in patients with β-thalassaemia major and advanced liver disease*. Br. J. Haematol. 2017; 178(1):130-136.

[42] Origa, R., Ponti, M.L., Filosa, A., *et al. Treatment of hepatitis C virus infection with direct-acting antiviral drugs is safe and effective in patients with hemoglobinopathies*. J American journal of hematology. 2017; 92(12):1349-1355.

[43] Hézode, C., Colombo, M., Bourlière, M., *et al. Elbasvir/grazoprevir for patients with hepatitis C virus infection and inherited blood disorders: a phase III study*. J Hepatology. 2017; 66(3):736-745.

[44] Sharara, A.I., Rustom, L.B.O., Marrache, M., *et al. Sofosbuvir/velpatasvir for chronic hepatitis C infection in patients with transfusion-dependent thalassemia*. J American journal of hematology. 2019; 94(2):E43-E45.

[45] Zamani, F., Ajdarkosh, H., Safarnezhad-Tameshkel, F., *et al. The effectiveness of sofosbuvir and daclatasvir in the treatment of hepatitis C in thalassaemia major patients and their effect on haematological factors*. J Indian journal of medical microbiology. 2018; 36(2):224-229.

[46] Maffei, L., Sorrentino, F., Caprari, P., *et al. HCV Infection in Thalassemia Syndromes and Hemoglobinopathies: New Perspectives*. J Frontiers in molecular biosciences. 2020; 7:7.

[47] Musharraf, S.G., Iqbal, A., Ansari, S.H., *et al. β-Thalassemia patients revealed a significant change of untargeted metabolites in comparison to healthy individuals.* Sci. Rep. 2017; (7,42249):1-10.

[48] Rower, J.E., Jimmerson, L.C., Chen, X., *et al. Validation and application of a liquid chromatography-tandem mass spectrometry method to determine the concentrations of sofosbuvir anabolites in cells.* Antimicrob. Agents Chemother. 2015; 59(12):7671-7679.

[49] Eipel, C., Abshagen, K. and Vollmar, B. *Regulation of hepatic blood flow: the hepatic arterial buffer response revisited.* World J. Gastroenterol. 2010; 16(48):6046-6057.

[50] Gregory M.Susla and J.L. Lertora, J., *Effect of Liver Disease on Pharmacokinetics*, in *Principles of Clinical Pharmacology*, 3rd Edition. American Society for Clinical Pharmacology and Therapeutics, 2013. Elsevier Inc. 81-96.

[51] Harvoni data sheet V8.0-(updated 13th of August 2019) NewZealand data sheet. 2019. Available from: https://www.medsafe.govt.nz/profs/datasheet/h/HarvoniTab.PDF. Accessed: 10th of June 2020.

[52] Eknoyan, G., Lameire, N., Eckardt, K., *et al. KDIGO 2012 clinical practice guideline for the evaluation and management of chronic kidney disease.* Kidney International Supplements. 2013; 3:19-62.

[53] Keating, G.M. *Ledipasvir/Sofosbuvir: a review of its use in chronic hepatitis C.* Drugs. 2015; 75(6):675-685.

[54] Geddawy, A., Ibrahim, Y.F., Elbahie, N.M., *et al. Direct acting anti-hepatitis C virus drugs: clinical pharmacology and future direction.* Journal of translational internal medicine. 2017; 5(1):8-17.

[55] Limenta, L.M., Jirasomprasert, T., Jittangprasert, P., *et al. Pharmacokinetics of deferiprone in patients with β-thalassaemia.* Clin. Pharmacokinet. 2011; 50(1):41-50.

[56] Iman A. El-Baraky, Maggie M. Abbassi, Fatma S. Ebied, *et al. Does Beta-Thalassemia Major alters Sofosbuvir/Ledipasvir exposure in Hepatitis-C virus Infected Adolescent Patients?* journal of Clinics and Research in Hepatology and Gastroenterology. 2021.